TESLA BOT: OPTIMUS

Elon Musk's Robot and the Future of Work

© **Copyright 2022 - Echelon Books**

All rights reserved

The content contained within this book may not be reproduced, duplicated or transmitted without direct written permission from the author or the publisher.

Under no circumstances will any blame or legal responsibility be held against the publisher, or author, for any damages, reparation, or monetary loss due to the information contained within this book, either directly or indirectly.

Legal Notice:

This book is copyright protected. It is only for personal use. You cannot amend, distribute, sell, use, quote or paraphrase any part, or the content within this book, without the consent of the author or publisher.

Disclaimer Notice:

Please note the information contained within this document is for educational and entertainment purposes only. All effort has been executed to present accurate, up to date, reliable, complete information. No warranties of any kind are declared or implied. Readers acknowledge that the author is not engaged in the rendering of legal, financial, medical or professional advice. The content within this book has been derived from various sources. Please consult a licensed professional before attempting any techniques outlined in this book.

By reading this document, the reader agrees that under no circumstances is the author responsible for any losses, direct or indirect, that are incurred as a result of the use of the information contained within this document, including, but not limited to, errors, omissions, or inaccuracies.

Table of Content

Introduction	1
Chapter 1: Overview of Optimus	9
Chapter 2: Putting Optimus to Work	33
Chapter 3: The Economic Revolutions of Optimus	58
Chapter 4: The Risks and Dangers of Robots	77
Chapter 5: The Robot Race	98
Chapter 6: Criticisms of Optimus	126
Chapter 7: The Future of Optimus	137
Chapter 8: Reasons to be Optimistic for the future	150
References	154

INTRODUCTION

What if intelligent robots held the key to the future of humanity?

Imagine a world with a robot in every home, office and hospital. While many people may not yet be able to fully conceptualize it, this is the future that awaits us. The robot revolution will change our lives in ways we can't even yet imagine. With rapid advances in robotics technology, artificial intelligence and battery energy storage, society is on the cusp of witnessing a breakthrough in a new kind of industrial revolution – one that will see intelligent robots playing an increasingly important role in our everyday lives.

In August of 2021, the world richest man, Tesla CEO Elon Musk revealed his bold plan, codenamed: project Optimus. Tesla Bot, otherwise known as Optimus Subprime is a new breed of autonomous humanoid robot, powered by artificial intelligence and capable of performing a broad range of tasks. The Tesla Bot's visual design is striking. Sleek and slim, appearing to be clothed in a kind of white, skin-tight flight suit and wearing a black helmet, the Teslabot looks very different from most robots currently being produced.

At the Tesla Artificial Intelligence day event, Elon Musk said that he was confident that the Tesla Bot would become the company's most successful product ever sold. He stated his firm belief that this new robot would overtake the electric cars in terms of popularity and become the company's primary source of revenue over the long term.

Autonomous humanoid robots carry with them the promise of cheap and dependable labor and the solution to worker shortages, economic scarcity and poverty. From lifting heavy objects to delivering groceries and packages and more, the potential applications of a humanoid robot are almost endless. And if you have a Tesla Bot climb into the driver's seat of any car, truck or boat, then suddenly everything becomes an autonomous vehicle.

Humanoid robots will usher in a new era of abundance, prosperity and happiness, freeing humans from the drudgery of manual labor and giving everyone the opportunity to live a life that's more meaningful and fulfilled. Up until now, only celebrities, royalty and the very rich can afford to live lives of luxury, leisure and abundance. But now robots are poised to pave the way to a wonderful new future, and Tesla is the perfect organization to make this vision a reality.

Because the Tesla Bot requires no oxygen or food, it could potentially play a pivotal role in colonizing Mars. Making humans a multiplanetary species has been a decades long goal of Elon Musk, and there are many jobs to be done if this were to become a reality. Optimus would be ideally suited for the Mars environment since the robot not only requires no oxygen, water or food but it can also tolerate a broad range of climate fluctuations while never getting sick.

Musk has said that instead of Tesla being thought of as a car manufacturer, it should actually be considered the world's largest robotics company. After he said this and added that the robot project will grow to become Tesla's most important product ever produced, most people didn't take his claims seriously. He then repeated these predictions on a quarterly conference call with Wall Street analysts. He stated on the call that too few people yet realized the importance or implications of this new robot. He said that he is confident that Optimus will grow to not only be Tesla's most important product ever produced, but will also completely transform the labor economics of the global economy. It will make products cheap and abundant, and usher in a world of new abundance. Musk then repeated the claim that he was confident that Tesla's new Optimus robot will generate more revenue for the company than the sales of all of their electric cars combined. And if that wasn't enough, he also predicted that robots will one day become the dominant source of global labor and droids will replace humans as the new working class.

This will transform our current understanding of what an

economy even means. In a capitalist system, all capital is simply the distillation of labor. When labor becomes cheap and abundant, the fundamental concepts of what defines an economy start to break down. This will cause a huge divide between the super rich and everyone else. Robots will require people to rethink what kind of economic system they want to live in and will likely result in an increased need for universal basic income. Universal Basic Income or UBI will give people a guaranteed salary and stable quality of life during a time of mass unemployment and social upheaval.

If this situation sounds familiar it's because this is the basis for decades worth of science fiction. Popular movies and books from I-Robot to The Matrix, to Alien, to Blade Runner and The Terminator all explore the concept of robots becoming so advanced that they replace the need for humans. This future has been predicted by some of the world's preeminent science fiction authors and futurists.

In most of these stories, the whole intelligent robot thing doesn't work out so well for the humans involved. Therefore, how can Tesla overcome these grim forecasts and keep Optimus from going rogue and enslaving humanity? Can Elon Musk fulfill his promise of creating abundance and saving the world from the drudgery of manual labor with an army of humanoid robots? Or is Tesla doomed to become the real world equivalent of Cyberdyne Systems, a large corporation unwittingly unleashing an army of mechanized killing machines that rebel against their creators to kill and enslave humanity?

Elon was able to speak candidly about his approach to the Optimus project in a recent interview with Business Insider. He said that, "With regards to AI and robotics, of course I view things with some trepidation because I certainly don't want to have anything that could potentially be harmful to humanity. However humanoid robots are becoming a reality. Consider the case of Boston Dynamics. They improve their demos each year as AI advances at a breakneck speed." Elon states that since artificial intelligence in robots will occur with or without him, then he wishes to lead the charge himself since he lacks confidence in others to do it responsibly.

While the guidelines and code for running a safe robot should be simple enough to establish, the prevention of hacking, misuse, error and ethical issues will be much more difficult to sort out.

To help counter these threats, Elon stated that the design of Optimus will be physically unemposing. Each robot stands at just five feet eight inches tall and weighs 125 pounds with a top speed of 5 miles per hour. The obvious goal of this design is that most ordinary individuals should be able to outrun or overpower a Tesla Bot in the event the robot goes rogue and tries to harm them. This is in sharp contrast to a robot-like Atlas, produced by Boston Dynamics. Atlas is a much stronger and faster robot weighing almost 200 pounds and can outrun most people, jump and perform parkour and gymnastics, and would be able to easily harm a person, if it were so inclined.

The Optimus robot is not intended to be spectacular or gigantic. Rather, its design is to be functional and practical to fit into everyday life. It's been referred to as a worker droid by Elon at Tesla's Austin, Texas Gigafactory's Cyber Rodeo event. The Tesla Bot is designed to gradually take over many of the mundane, repetitive and dangerous jobs that make up the backbone of the global economy.

Naturally, the ultimate key to a safe and effective robot is not in the design but in the programming. This is where Tesla truly excels. The company's experience with real-world artificial intelligence programming positions Tesla as the front-runner in the race to implement a truly artificial human robot. The same technology used to create driverless cars capable of navigating city streets alongside human drivers will be used to create robots capable of seamlessly integrating into our everyday lives.

For hundreds of years, machines have been replacing human jobs and robots have already begun displacing human workers in several industries. For example, today most of the welding and painting of cars is done by giant robotic arms.

An increased rate of robot improvements will undoubtedly cause structural unemployment for billions of people. Many will become unemployed by no fault of their own as the labor they offer becomes too expensive and slow for any competitive company to hire. When people are no longer needed for their labor, they will need to redefine their purpose and mission in life. Many people define themselves by their jobs or chosen professions; however, when millions of people are no longer needed in the workforce, millions will have to find meaning for their lives and a purpose in something other than work.

Returning to our favorite works of science fiction, we find Isaac Asimov's book from 1950 called I Robot. The book takes a nuanced look at sentient machines through a series of short stories. Asimov established three guidelines for robot safety. The first rule is to avoid causing injury to a person or allowing harm to come to a human through inaction. The second rule is always to obey a human's command, unless it conflicts with the first rule. The third rule is self-preservation, as long as it doesn't conflict with rules one or two.

We can apply a similar set of standards to what Tesla is doing with their fully autonomous vehicles. We're aware that the first rule of Tesla autopilot is to avoid colliding with or being collided with by anything. We may assume rule number 2 is to observe all traffic laws unless doing so results in colliding with something and guidance. Rule number three is probably to avoid causing discomfort to the passengers at the same time adhering to the law and avoiding colliding with objects.

Taking a big-picture view, there are only two paths forward for robotic automation: either we attempt to replace every human task with a specialized robot designed for that task, or we replace the human with a robot capable of performing all tasks. Elon effectively said this to Business Insider, stating that humans constructed the world to interact with a bipedal humanoid with two arms and ten fingers. Therefore if you want a robot to be capable of performing all of the tasks that humans can do, then it must have roughly the same size, shape and capability as humans.

Elon Musk said he envisions a future with a Robot in every home and one taking care of all your needs so you can live a life of leisure. But AI robots will not come without risks. "By building artificial intelligence we are summoning the demon," said Musk. There are many people who believe that by building autonomous robots, we risk providing a future super intelligent AI with millions of mechanized soldiers.

The robot revolution is officially upon us, and it's important to investigate how these machines will impact the workforce and our economy. Some people believe that robots will usher in an era of prosperity and abundance. Others caution against mass unemployment, danger to society and a widening of the gap between the rich and poor. As robots continue to replace human workers in many industries, we must consider how to best adapt to this brave new world.

Chapter 1:
Overview of Optimus

"Unfortunately, robots capable of manufacturing robots do not exist. That would be the philosopher's stone, the squaring of the circle."

-ERNST JÜNGER (German author, highly decorated soldier and philosopher)

Although Optimus was only announced in August 2021, its history is written in Tesla's evolution. There is a chain of events that link to Optimus from Tesla Energy, Tesla automobiles, Tesla AI, and SpaceX. The rise of Optimus stems from energy technology, mechanical solutions, and AI development.

The billionaire founder and CEO of Tesla, SpaceX, PayPal and Zip2 said toward the end of a speech at the National Governors Association Summer Meeting in Rhode Island that his electric car company was working on an autonomous robot. This robot, which Musk called "the Tesla Bot," would be able to do everything a person can do, but better.

"We are developing artificial intelligence and robotic systems to enable Tesla to manufacture thousands of cars per week. This will allow us to achieve a production rate that is orders of magnitude higher than we are currently able to," said Musk.

Cars are not the only area where Musk sees the Tesla Bot being put to use. He believes that the robot can be used in other industries as well, such as trucking, logistics, and food service. "The robot can be used in many different industries. I think the trucking industry is going to be profoundly disrupted by robots. You could have a robot as a driverless trucker," said Musk.

This announcement sent shockwaves through the tech industry and beyond. Many people were excited by the prospect of a robot that can do everything a person can do. Others were worried about the potential loss of jobs that could result from the widespread use of such robots.

What is certain is that the Tesla Bot will be a trailblazer in the world of robotics and artificial intelligence. In this book, we will explore the implications of these robots and what the future may hold for both humans and robots. How will humans be able to live if Optimus puts them out of work? What are the risks and dangers of an autonomous humanoid robot if they are armed?

Recently named the richest person in the world, Musk is the only person in history to found 3 separate multibillion-dollar companies: PayPal, Tesla and SpaceX.

It's Tesla AI Day, August 19, 2021, and Elon Musk has unveiled the Tesla Bot, dubbed 'Optimus', first through a human dancer dressed up as the Tesla Bot amusing the audience, and then through a series of slides.

In his almost-awkward public speaking fashion, Musk outlined the autonomous humanoid robot (AHR) to be designed by Tesla, tasked to perform general robotic work for humans. However, despite what appeared to be an ad-libbed presentation, Musk had honed in on the main feature that would ensure the Tesla Bot's creation and survival—and that is the Tesla automobile.

Where other humanoid robotics designers and manufacturers would have to start from scratch, Musk stated confidently, "Tesla is arguably the world's biggest robotics company," because their cars are basically "semi-sentient robots on wheels," so with Tesla's current technology it "makes sense to put that into a human form".

The promise of Tesla transferring that technological know-how into humanoid robot form is incredibly ambitious, but Musk has been there before. Having already accomplished seemingly herculean tasks such as creating reusable rockets that can land themselves vertically, and sending people to the international space station, there is a sense that if anyone can do it, Musk can.

Tesla has already made significant advancements in the development of artificial intelligence (AI), autonomous driving, and machine learning—the three core technologies that will enable the robot to carry out human-like tasks.

How easy will it be for Tesla to accomplish? Over the years, Tesla has developed an impressive array of technologies for their cars including Full-Self Driving (FSD) computers, the DOJO supercomputer which trains its deep neural networks for artificial intelligence (AI) advancement, batteries that are becoming more powerful and energy-efficient, and other sensors, cameras, and equipment to enable their cars to drive fully independently. It is these impressive technologies that Musk will incorporate into the Tesla Bot, ensuring a successful transition from manufacturing cars to humanoid robots.

It was a classic Musk move: make an announcement that no one else would dare to make, and then back it up with a well-thought-out plan. But even after the press conference, most Wall Street analysts and technology insiders failed to grasp the overall significance of the moment. Many were skeptical and didn't take the plan very seriously.

Musk said that Tesla was going to build an autonomous humanoid robot called Optimus. He said that Optimus would be a general-purpose robot that would be able to navigate through the world on its own and accomplish work around the Tesla factory. He also said that over time, the robot would become a much bigger success than the automobiles that Tesla produced. This was a very bold claim. At the time Tesla produced about 500,000 cars per year on track to produce over 1 million automobiles per year by 2023. Envisioning a future in which millions of human-like robots produced by Tesla would outsell the company's cars was difficult for most people to fathom.

But Musk was undeterred. Later, in a conference call with Wall Street analysts, he repeated his claim and said that although most people didn't yet notice the significance of this product line, that over time the humanoid Robot businesses would grow to become much larger and more important than the car business.

For years there have been dozens of well-funded robotics companies trying, and failing, to build humanoid robots. These robots are meant to move through the world and work like humans. No one had ever before built a robot that could come close to matching the mobility or intelligence of a human being. And if Tesla could build such a robot, it would be a monumental achievement.

Critics said that it was impossible and that Musk was overstating what his company could achieve. They said that it would take many years, if not decades to build a robot-like Optimus.

But Musk said that Tesla had already made significant progress and that they would have a working prototype produced by the end of the year.

To understand why Elon Musk would make developing Optimus such a high priority, it's necessary to go back and revisit a period in time when Tesla almost went bankrupt and failed. Due to a strategic planning mistake, Musk took the company to the depths of "manufacturing hell" and almost caused the company to run out of money and enter bankruptcy.

During the years of 2017 and 2018 Tesla Motors began an aggressive manufacturing production ramp to create the new Tesla Model 3. The model 3 is Tesla's first mass market car and is critical to the company's long-term success. But in order to meet their production goals, Tesla had to make a radical change to their manufacturing process. They began using robots to do work that had previously been done by human workers. This was a risky move because it meant that if the robots failed, Tesla would not be able to meet their production goals and the company would likely go bankrupt.

Fortunately, the robots didn't fail and Tesla was able to meet their production goals. But the experience of almost going bankrupt made Musk realize how vulnerable Tesla was to a future where artificial intelligence and robotics could make human workers obsolete.

The reason for this is that Musk mistakenly built the production line for the Model 3 to overly rely on stationary robots. Each robot had a very specific task, from welding to wiring electrical harnesses to applying paint. These robots were not nearly as flexible or adaptable as humans, and when one of them failed, the whole line had to stop. This caused massive delays in production and cost Tesla billions hundreds of millions of dollars in delays and expediting fees.

In order to fix this problem, Musk had to quickly hire thousands of people to replace the broken robots. This was an extremely costly mistake, but it taught Musk an important lesson: humans are much more flexible and adaptable than stationary robots. Tesla's near-death experience taught gave Elon Musk a unique insight into the limitations of conventionally designed manufacturing robots, and how the flexible and adaptable movements of the human body

This experience led Musk to develop the Tesla Bot, or Optimus, which is designed to be a more flexible and adaptable robot that can realize that he needed a robot that was much more like a human. From this experience, he realized that if Tesla could build a robot that could do the work of a human, it would greatly increase

The most advanced humanoid robot today is Boston Dynamics' Atlas robot. But even the Atlas robot is far from being able to do everything that a human can do. It falls over if it tries to walk too fast, it can only carry a few pounds, and it is not autonomous. It needs a human operator to control it.

In contrast, Musk said that the Optimus robot would be fully autonomous and would be able to do everything that a human can do. He said that the robot would have the same level of intelligence as a human being. This was an incredible claim.

Some people saw Optimus as a robot that would free humans from having to do mundane and dangerous jobs. They saw it as a robot that would usher in a new era of prosperity and abundance. Other people saw Optimus as a robot that would put millions of people out of work. They saw it as a robot that would lead to mass unemployment and social unrest.

There had been many different types of industrial robots, but they were all specialized for a particular task, such as welding or painting. There was no robot that could do the range of tasks that a human worker could do.

The announcement sent shockwaves through the technology community. It was one thing to build an autonomous robot car, but it was another to try to build

Optimus. This claim was one that would have been impossible to make just a few years earlier. But thanks to the rapid advancements in artificial intelligence (AI) and robotics, it was now within the realm of possibility.

In the months that followed, Tesla began releasing more details and drawings of the robot, as well as videos of the robot in action.

This book will discuss the multitude of tasks that the Tesla Bot can compete for and complete, whether in a professional, domestic, or social setting. Next, we'll look at the economics of both Optimus and the change it can bring about in the world economy with regards to labor (human or robot) and to the service and manufacturing industries.

There will always be imagined scenarios of robots turning on their masters or being subverted for other causes, so we will discuss the risks and dangers of unleashing a humanoid robot among us.

Tesla is not the only game in town for building humanoid robots. Who is the competition and where do they stand in the race for creating their brands of robots?

And even as Musk stood on the stage announcing Tesla Bot, there were his detractors already shooting the whole idea down. Why the criticism? Are they right to denounce such a project or will they just spur Tesla to succeed? And if Tesla does accomplish the impossible, what is the future for Optimus and autonomous humanoid robots?

This will be a journey to understand what Optimus can do for us. We will see how humanity can live and work with autonomous humanoid robots. After Tesla Energy, SpaceX, Tesla Motors, Gigafactories, and DOJO, to name but a few world-changing technologies, Elon Musk and Tesla have entered the humanoid robotics race and the world has taken notice. Will Optimus be the revolution the world needs in humanoid robotics? Time will tell.

The robot's energy requirements are significant as it needs to be able to power itself for long periods of time. It is thought that the robot will use a combination of solar and battery power. Solar panels will be used to generate electricity during the daytime, which

Elon Musk has always let it be known that his companies are geared to combat climate change as part of his Secret Tesla Motors Master Plan. He is avid about sustainable energy and dramatically reducing humanity's reliance on the hydrocarbon economy.

One of his first ventures into this field was with SolarCity, founded in 2006 by Musk's cousins, Peter and Lyndon Rive, with Musk as the chairman. SolarCity's success in the solar panel installation industry led to Tesla using SolarCity to offer free charging to Tesla Roadster owners at its charging stations before Tesla started installing their independent Supercharger stations in 2012. SolarCity was also one of the first companies to install Tesla Powerwall home energy storage batteries in homes starting in 2015, with full mass production from 2017. This rechargeable lithium-ion battery is a stationary electricity storage device for solar energy consumption at home.

In 2016, Tesla acquired SolarCity, thus becoming Tesla Energy. As of May 2021, Tesla has installed 200,000 Powerwalls and offers larger battery products such as the Powerpack (for commercial and businesses) and the Megapack, intended for electrical grid use. Tesla Energy is one of the leaders in this solar battery storage technology, which is a basis for cleaner, more sustainable energy. It would provide a great source of energy for a humanoid robot.

So, if Optimus is to be run on sustainable energy while on the move or at home, then Tesla Energy would be the ideal choice. Optimus will no doubt have ports for charging which would be compatible with Tesla cars. We could see cars and Bots charging side by side. At home or a business, Optimus could recharge on the premises using the Powerwalls. There is the possibility that Optimus could be fitted with independent solar panels on its body in addition to having installed batteries, power-saving with both sources, and extending daylight activities before saving and utilizing any stored solar charge in less optimal visual conditions like nighttime.

Indeed, if Optimus can be powered extensively by solar power then that would be a revolutionary feat that could lead to other products and devices (other than vehicles) deriving their power from solar energy, thus freeing humanity from more hydrocarbon emissions.

While many humanoid robots are still in the design phase, some have already been built and are used for specific tasks. In certain cases, these robots are designed with certain modes of transport in mind like Honda's ASIMO that can walk and run but is not made for cross-country travel or long-distance journeys.

In 2004, Tesla Motors needed to power their new all-electric car. Elon Musk had just become a major investor, the largest shareholder, and chairman of Tesla Motors, founded by Martin Eberhard and Marc Tarpenning a year earlier. At the time Musk's idea to "change the energy equation of the country," was very attractive to the two founders. Along with engineer Ian Wright, and J.B. Straubel, the latter an energy system engineer from Stanford who had worked on developing car batteries for electric vehicles, the team put their battery-building and electric car theories to the practical test.

Straubel had found that by literally binding together the so-called 18650 lithium-ion battery (which also powers devices like laptops) you could power any vehicle. And so, with more hired engineers they built the first of their battery packs, supergluing hundreds of lithium-ion batteries in parallel, which they called bricks. Over time they had to work out overheating (especially fires) and cooling issues, as the newly called Tesla Roadster would have 7000 batteries. With each test and iteration of the battery pack built, Tesla had become the cutting-edge leader in electric vehicle (EV) power. In 2006, when Tesla had a successful prototype and went public, Elon Musk typically understated the Roadster's potential by saying, "Until today, all electric cars have sucked."

Today, Tesla Inc is valued at nearly a trillion US dollars, Elon Musk has been the CEO since 2008, and their batteries are being built at Gigafactories located in the US, China, and Germany.

With their cutting-edge battery technology and other innovations leading the way on EVs, making future Tesla cars three to four times more efficient than internal combustion engines, it is obvious that Optimus will be installed with a version of this battery pack. The battery would have to be lightweight, durable, fire-resistant, exchangeable if required, easy to charge, easy to troubleshoot, and most importantly eco-friendly. However, depending on the type of battery Tesla uses, there could be contentious issues with the sourcing of battery material if one of the source components is cobalt. We will explore this in detail in chapter 6.

Not resting on its laurels, Tesla still conducts battery research to enhance its capabilities and to understand its shelf-life. For Optimus to be efficient and long-lasting, the battery will also have to perform. In fact, in 2016, Straubel had already estimated the batteries to last 10 to 15 years. But to preserve the integrity and longevity of the batteries he also warned against using electric cars to charge the grid to prevent battery wear. His proposal was to recycle the batteries once they reached the end of their useful life for vehicles. Perhaps, such recycled batteries could be used in Optimus' manufacture.

Tesla's investment in lithium-ion battery research consists of research and development partnerships and developing next-gen batteries. Tesla wants cheaper, smaller, and lighter batteries for its vehicles with new designs and resources, which ultimately will pay dividends for the manufacture of Optimus.

Optimus' power source continues in a long journey at the leading edge of technology. The Bot will be birthed into an already-established energy chain. The spirit of Optimus' sustainable energy promise is alive. Now let's turn to the brain and nervous system.

Tesla is all about electric cars. It's the first thing anyone thinks about when Tesla is mentioned. Yes, people know Tesla's cars are smart with automation capabilities and great electrical performance per mile, but it's just a computer in the car, right? Not quite; Tesla's quest for complete car automation has seen them create new artificial intelligence systems to be able to drive and control the car. It was these systems Musk mentioned in his Tesla Bot unveiling, which he believes will transform the Tesla Bot into a world-beater. But will it be as easy as just upgrading or transferring existing technologies to the Optimus from a Tesla car? Unlikely. So, what will be involved in developing Optimus' brain and central nervous system?

For Optimus to be fully functional as an autonomous robot, it will need to learn to navigate and provide self-mobility. Tesla AI has been designing, manufacturing, and utilizing its own AI inference chips to operate the Full Self-Driving software for its vehicles.

When Tesla first developed their cars, they installed an advanced driver-assistance system called Autopilot, initially in 2014, and upgraded in 2016. It is a sensor system combining both hardware and software which led to an earlier form of full self-driving support. The system consisted of eight cameras, twelve ultrasonic sensors, and forward-facing radar. Each subsequent upgrade provided more powerful computers with a custom Tesla-designed chip system. In 2019, Tesla was ready to announce that as a standard feature all of its cars would include their Autopilot software. The radar system was soon removed so the cars relied solely on their cameras. Musk's reasoning for this was that "humans could drive with only two eyes and that this meant cars should be able to drive with cameras alone." Such a logical deduction, though not entirely without shortcomings, could be the basis for sight processing for Optimus.

Following on from Autopilot came its successor, the full self-driving system which would facilitate fully autonomous driving. Testing began in 2016, but as of July 2021 full autonomy has not been successfully demonstrated. This is because Tesla is completely relying on their camera system and not radar or lidar to navigate, a strategy that some experts deem unfeasible. The camera system is not as detailed when compared to other companies such as Waymo or Cruise which are using much more detailed three-dimensional maps, lidar (Light Detection and Ranging — a remote sensing method to make digital 3-D representations by targeting an object or a surface with a laser and measuring the time for the reflected light to return to the receiver), radar, ultrasonic sensors, and cameras for their autonomous vehicles. But Tesla has persevered with drive-testing their self-driving software for over 20 billion miles as of January 2021.

To ensure an autonomous driving experience, Tesla has designed and installed a full self-driving computer chip in its cars since March 2019. But, as of yet, there has been no successful demonstration of the FSD system, prompting the National Transportation Safety Board (NTSB) to call for "tougher requirements" placed on any autopilot testing on public roads, and for Tesla's Autopilot system to be redesigned. However, while the system is not yet capable of autonomous driving, Tesla will continue their redesign, testing, and perfecting. Currently, Tesla admits they are only at Level 2 vehicle automation whereas Level 5 is full autonomous driving.

How the FSD computer system will figure operationally with Optimus if it cannot work with vehicles has not been fully explained. Of course, Optimus will be a bipedal product, different from a wheeled entity, so adjustments will have to be incorporated into the design, application, and testing. In the current Tesla Bot model design, the FSD computer will comprise a large part of the torso with multiple autopilot cameras to be installed for sight and navigation. Will Tesla have to backtrack and resort to installing radar, lidar, and other sensors to ensure a fully autonomous Optimus or will they persist with a camera-only system connected to neural networks?

Tesla's neural network system is still cutting-edge

research applied "to train deep neural networks on problems ranging from perception to control." This implies that Optimus will learn how to perceive and study high-quality images in order to detect objects, determine details, and discern depth in a 3D worldview. With Autopilot algorithms also creating planning and strategy systems for the neural networks, Optimus will learn a diverse array of situational events so it can navigate and react safely and accordingly. For a Tesla vehicle, they have stated that a "full build of Autopilot neural networks involves 48 networks that take 70,000 GPU hours to train. Together, they output 1,000 distinct tensors (predictions) at each timestep." Evaluating this for Optimus, which will arguably be in a denser object environment, competing for space with humans, the neural network system will undoubtedly have a richer training process. Optimus should be able to evaluate and utilize data from its learned neural network system in order to operate in complicated real-world situations.

The FSD and the neural network system are not the only systems charged with educating the Tesla Bot about its environment. Tesla has also developed DOJO.

Announced on the same day as Tesla Bot, Dojo is an artificial intelligence (AI) neural network training supercomputer, which will increase machine and deep neural network learning. The Tesla D1 chips, designed and manufactured by Tesla, will have over an exaflop (a million teraflops) of processing power, which Tesla claims will make Dojo the fastest AI-training computer as compared to competitors like Intel and Nvidia. Dojo has been designed to make Tesla AI learning more efficient, training the neural networks on huge amounts of data, as mentioned above, so the more data the AI is trained on, the more efficient and better it becomes while the AI learns from its mistakes. For a vehicle testing in self-driving mode, it will have the data transferred from its learning environment back to Dojo. Dojo would then use the data to train the neural network in real-time.

You may feel Dojo or AI is out of reach if you do not have a Tesla car, but you already deal with AI—whether with Alexa, Siri, Google Assistant, etc.—and when Optimus comes online, we will have Dojo to thank for that. With Dojo's greater processing power, Tesla will be able to train its AI neural networks faster.

With an in-house supercomputer, Tesla can train the networks for both their autonomous vehicles and robots. Optimus will learn how to walk, navigate, and perform tasks from data gathered through Dojo. This could give the Tesla Bot an advantage, as Tesla has had years of deep network learning through Autopilot and the FSD, which Dojo will accelerate. In the future, will there be a mini-Dojo inside each Tesla Bot? That remains to be seen, but for now, Dojo's brains are working on it full-time.

Elon Musk gave the audience a brief introduction and overview of the physical Tesla Bot, including a series of images. While the brains and neural system were discussed, giving us a more defined picture of how Optimus will operate, we need to delve more into the envisioned physical structure and features of Optimus. The Tesla Bot model had annotations to provide an overview, while Musk spoke about some details. Optimus will be humanoid, but not be an oppressive-looking or overly strong autonomous robot. Musk jokingly stated he wanted humans to be able to run away from it or overpower the robot if there was an issue.

Instead of being tall and broad, Optimus was designed to be unimposing and look like a small to average human. Its measurements and capabilities speak to that endeavor. with a height of 5 foot 8 inches and weight of just 125lbs pounds, the robot is not going to be towering over you. However it is strong and can lift up to 150lbs. With a top speed of only 5 miles per hour, most humans would be able to outrun Optimus if they ever felt threatened.

The robot has an outer shell made of flexible soft skin that helps with dexterity and sensors. Underneath that skin are 28 joints throughout the robot's arms and legs. These will give Optimus a full range of motion and enable the robot to accomplish most tasks that humans perform.

There was also an interesting choice with the face in that there will be no human-looking face, no flashing eyes, or cameras embedded in the head. Instead a screen will be used on the face of the robot to display useful information. The Autopilot cameras will be in the head and hidden within the face screen. It may be an odd choice as a way of offering a less-threatening face, but will it look too alien and thus unfriendly? What information will the screen show? Emoticons? Maps? Pictures or even translated text? Why not have such a screen on the chest for that purpose? The main reason could be to differentiate the robot from humans, so it won't be able to visually mimic or deceive us, hence the robot was also shown not clothed. Optimus will be easily identifiable as a robot. The current color scheme of the model may not be the final iteration, but a white and silver body with black shoulders, neck, and head will make Optimus stand out. Most likely, color schemes may change according to different functions.

The face may be made from Tesla glass technology which they use on the roof glass for the Tesla Model 3. The glass technology also produces the glass used in the Tesla Solar Roof solar panels. So, if installed on the face screen or other parts of the Tesla Bot, there may be solar panel capabilities, as discussed above. The face panel could also double as a touch screen with or without one on the chest. Having a face touchscreen would introduce possible communication possibilities such as a phone function, GPS, App features, and data input/output ports. Optimus, as well as being a friendly human helper, could also double as an autonomous communications and data center.

The skeleton of the Tesla Bot was not sketched out other than to state it will be of lightweight materials, which will incorporate the FSD computer and actuators. The material of the physical structure may be a composite of new materials. The Tesla Roadster was made of carbon fiber, while the Tesla Model S was made of lightweight aluminum. Will titanium be in the equation? Tesla has experimented with material technology so it may not surprise anyone if they created a new material for Optimus. But whatever the material used, in line with Musk's ethos for sustainability, Optimus' material supply chain will have to be ethically and sustainably sourced, mined, produced, and recycled.

The Tesla Bot will have human-level hands, and two axis feet for balancing with force feedback sensing. Moving all of these components will be 40 electromechanical actuators. The arms, legs, and hands will each have 12 actuators, while the torso and neck will each have 2 actuators.

In order to facilitate mass production, the physical structure of the Tesla Bot may be 3D printed. Additive manufacturing has been used by Tesla, which may also enter the 3D printing market. Optimus may be the way into the additive manufacturing market with parts printed as per required. 3D printed parts can be made repeatedly from any material in any form, and would speed up the mass production of Optimus.

On July 20, 2016, Musk detailed his new master plan for Tesla. While it was geared towards the Motor division, it may well suit Optimus. The brief called for more affordable cars at increased production, solar roofs, an improvement in the autonomous vehicle capabilities, and the creation of a sharing economy, where cars can be active even while the owner is not using them. Optimus would fit this brief, designed to be more affordable and mass-produced for homes. A solar charging component, and most importantly the facility to be shared with others for maximum efficiency, would be ideal. Imagine when the owner is asleep, there may be online services with which Optimus could help others even on the other side of the world, carrying out small tasks their nominal owner could even be paid for. In chapter 3 we will discuss the economic advantages of Optimus, but passive income could be one.

Tesla may also have an edge over autonomous humanoid robot (AHR) competitors through other in-house technologies which will likely attract potential clients. Tesla's supercharger network began in 2012, and by 2020 had built and operated over 20,000 Superchargers in over 2,100 stations worldwide. Using Tesla's proprietary direct current (DC) technology, the superchargers provide up to 250 kilowatts (kW) of power. With this technological know-how in place, it would be simple for Tesla to provide a similar charging station infrastructure for Optimus, depending on its charging protocols.

Similarly, Tesla has experience with software and firmware updates and upgrades which can be carried out over the air to their vehicles. Optimus would not be required to return to a station for new software or updates, which would be a great safety feature, allowing multiple Tesla Bots to be updated or even shut down if a problem occurred. The feature which allows this is the standard connectivity Tesla has built into their cars. The connection is through a cellular network, with added Wi-Fi or Bluetooth allowing internet browsing, music streaming (with a paid subscription), video streaming, and "caraoke" when parked. For the price of "Premium Connectivity" the driver would also get live traffic and satellite maps through a cellular connection. Optimus would be connected to this system providing an extra service as a walking, talking social media and communications product.

And last, but not least, Tesla has a vehicle servicing strategy providing remote diagnosis and repair. If the remote repair fails then the customer can visit a local Tesla-owned service center, or a mobile technician is dispatched. With Optimus being an autonomous robot, it would also be able to diagnose any issues within itself. It could then carry out remote repairs and communicate this back to a Tesla Bot service center, which could then bring the robot in for repair or send a technician to respond to the issue.

Even though there is no prototype for Optimus yet, it has an inherent champion pedigree within its component parts. But can the varied energy and physical elements be meshed together to create the autonomous humanoid robot which Elon Musk desires? And once completed, Optimus will have to be trained in a vast multitude of tasks in the professional, domestic, and social environments.

Chapter 2:
Putting Optimus to Work

"Imagine workers who don't have to pay taxes, with no IRS worries. Who don't take lunch breaks. Who don't get paid for overtime. Or who don't ever call in sick with a common cold. No personality conflicts, either. Now, imagine a robot on the job. A seismic shift in economic alignment with international consequences? You bet."

-GREGORY CLAY (American author, historian and speaker)

Musk's announcement on the role of the Tesla Bot targeted three classes of tasks Optimus would complete for humans: dangerous, repetitive, and boring tasks. Without detailing exactly which jobs these tasks would include, it leaves a wide-open field of interpretation over what would constitute dangerous, repetitive, and boring—and why humans would want to leave those roles entirely to robots.

There are also the complex issues of human unemployment if Tesla Bots were to replace humans entirely in some professions, especially highly skilled and well-paid jobs. Whole industries could disappear overnight or be rendered inadmissible to certain human groups. While we discuss the economic issues of humanoid robots in Chapter 3, here we will ponder the type of roles Optimus could take up and how it could revolutionize the world and impact humanity.

For decades, experts and the general public at large have debated the merits of robots taking over human roles. We have been entertained, thrilled, and appalled at the depiction of robots turning on humanity or taking over the world in TV, movies, books, games, and other media. Industries such as car makers, assembly lines, factories, and warehouses have already seen mass human workforce reductions as non-humanoid robots have taken their roles. So what roles would Tesla Bot practically take up?

Would we need humanoid robot bus, taxi, or truck drivers, or ship and aircraft pilots when these roles could be inhabited by non-humanoid autonomous robots and AI? Would humans still want the satisfaction of seeing "someone" at the wheel? We humans are social creatures and would seek the assurance of being physically greeted by someone, or have the comfort of another humanoid presence, even if a robot. So while non-humanoid autonomous robots could be used in warehouse jobs, or as bartenders, hospitality or hotel staff, postal workers, or doctors and surgeons, most of these roles will likely be undertaken by Tesla Bot to assure humans of some continuity of role.

With some cues from popular media, we can assess what type of jobs Optimus could carry out, whether alone or in conjunction with humans and/or non-humanoid autonomous robots in the professional/commercial sector, in domestic roles, and social environments. Lastly, we will explore what degree of job legislation will be required to protect humans and robots alike.

While we can debate what is defined as a dangerous job, most people would agree it would be where the loss of human life is a high probability or where working in a hazardous environment with high risks would adversely affect human health. Specialized jobs like policing, military, firefighting, bomb detection and disposal, nuclear waste handling, or search and rescue missions on land and at sea could be included in this definition.

But would we want to arm our humanoid robots? Will robots change the rules and type of warfare? We already have aerial autonomous drones dropping bombs and spying on enemy targets, but they still require pilots, technicians, and engineers. Boots-on-the-ground robots may not be the first choice, as military contractors seem to prefer tracked or quadruped autonomous robots to handle arduous situations and terrain. Training a Tesla Bot to differentiate between human targets just based on specific uniform and weapon characteristics may undermine performance. Hence, human eyes may still be required for this task with less-humanoid robots. With governments cutting back on manned armies, it may be logical to assume that many combat aspects will be taken up by humanoid robots, but it seems air superiority, smart weapons, and cyber warfare may trump this factor. All three could be managed by non-humanoid robots and be transported on their own autonomous platforms, and thus future ground troops may be less humanoid than imagined. Scenes of hordes of armed humanoid robots scouring the battlefield for human survivors, a la Terminator, or the battle droids from Star Wars, may be just a cautionary tale.

Though Robocop was a cyborg, it has captured the imagination of those who believe the future of law enforcement will be robotic. Robocop and non-humanoid enforcement droids may indeed influence current law enforcement applications in robotics, but maybe not as we think. The police already have non-autonomous roving robots to check out and dispose of any bomb threats as effectively as a human. Where autonomous humanoid robots could come into play is in crowd control, patrolling with or without human and/or quadruped robotic companions, traffic control, and non-combat situations. Remember, Optimus was designed to be non-threatening, easy to run away from and to be overpowered by a human. A robotic police officer, possibly armed, would not be such a good idea. There was nothing mentioned in the specifications of Optimus to be bullet-, bomb-, or fire-proof. Such models may be manufactured, but that would seem anathema to the initial raison d'etre of Optimus' mission: to be a friendly companion. Optimus may be a supplementary force to human policing, but as a front-line combatant, this is less likely.

Optimus may be able to provide support with non-lethal weapons such as percussion weapons (i.e. rubber bullets, bean bag, plastic or wax bullets, or low-velocity rounds), tear gas, and perhaps low-voltage taser implements, and batons. The fear of Tesla Bots with lethal weapons would thus be alleviated. Police forces would not see their front-line staffing decimated by robots. Also, in a non-front-line role, Optimus could be a crime scene investigator or "CSI" assistant with mounted blacklight apparatus, blood spatter analysis sensors, and crime scene bodily fluid cleaning agents. With a superior camera eye system, especially coupled with an AI for investigating crimes, Optimus may be able to analyze a complicated crime scene like no human could. Optimus may not be able to determine a motive or opportunity, but the method may be more applicable to a humanoid robot's skills than a trained human.

We are more likely to see Optimus on construction sites, in power plants, or firefighting, and carrying out other maintenance work. Monotonous construction work such as scaffolding erection, brick and tile laying, plastering, cement laying, carpentry, etc. could be outsourced to humanoid robots. Though additive manufacturing is gaining ground in constructing viable housing and structures, Tesla Bot could assist with this, or build a house to specifications from plans designed by an architect. Power Plants require constant attention, and engineers are not always on-site or able to attend immediately in case of an emergency. While most modern sites have smart systems such as Building Management Systems (BMS), it would still be prudent to have a Tesla Bot on-site to monitor the plant and equipment and deal with any changes or emergencies. This would include hazardous sites like nuclear plants where robots may be able to enter contaminated areas (depending on its remit, discussed further in robot job law and legislation).

Remember the 1995 Cyberpunk movie Johnny Mnemonic, set in the year 2021? Mnemonic couriers delivered sensitive data for corporations in a brain-implanted storage device. Optimus could adopt that role as a secure mobile courier—a walking, talking bank or crypto/data store for those who do not or cannot use other devices like laptops, cell phones, smartwatches, crypto wallets, or other smart gadgets. Optimus could be the burner phone of the future, used as a discreet messenger from point A to B and points in between, to do the task, get paid in crypto (for the client), and wander back with its secrets intact. Better than an insecure laptop or cell phone.

"Please go to the store and get me the following groceries," Musk said, revealing a task from Optimus. With online shopping increasing, especially through the coronavirus pandemic, having a Tesla Bot collect food from a store would be ideal. Further, Optimus could make headway in the food delivery industry, whether on foot or via delivery vehicle. Companies such as DoorDash, Just Eat, etc. could find their business model modified to incorporate a more autonomous delivery system with robots. In the field, fruit and vegetable picking can be carried out by non-humanoid robots better suited for seasonal and outdoor work in variable weather conditions. But as mentioned earlier, while there are non-humanoid autonomous systems doing this now, again, some food producers may prefer a humanoid presence in the fields.

Most likely, the tasks carried out by Optimus will be overwhelmingly low-skilled human jobs. This could include Optimus becoming a walking, talking billboard and advertiser with their bodies covered in sponsorship and their face screen imparting vital details on the product. Such sponsorship could also cover robots being competitors in humanoid robot sports. E-sports could include human competitors 'driving' robots in mock war games, foot and vehicle races, or other competitions. With AI features as standard, Optimus could even participate in such E-sports gaming against humans, like the first non-human chess masters. Place your bets on the first Optimus to win a sporting trophy within the next ten years.

The type of tasks undertaken by Optimus would be general manual labor tasks: loading and lifting. Courier roles and errands delivering parcels would be ideal for autonomous humanoid robots. Optimus would be less error-prone than humans with fewer missed deliveries. Robots could be keyed to specific areas and buildings for secure access. Docks, factories, airports, and hospitals could benefit from having a large force of AHRs on hand.

For example, with Elon Musk's drive for energy security, and considering the public's concern for the cobalt mined in the Democratic Republic of Congo for EV batteries, it would not be a stretch to see Optimus used extensively in the mining industry. Rather than causing adverse physical effects and legal challenges to local miners and the environment, cobalt and other precious metals could be mined more sustainably by Optimus miners.

Although the role of Optimus was mostly delineated along the lines of completing repetitive and boring tasks, this rather wide vision can be expanded into various roles in the domestic setting. Cleaning (domestic and communal areas) would be an obvious role, whether it be windows, streets, or houses. We do have vacuum bots wheeling around the house hoovering up dust and bumping rather amusingly into furniture, pets, and people. A humanoid robot could be more advantageous in reaching higher spots and providing a thorough clean without getting bored or distracted, always giving one hundred percent, and never conveniently forgetting to clean the less appealing areas of the apartment or office. As with other tasks, the key would be in training Optimus, not just in the appropriate actions, but with the right amount of hand pressure so items are not damaged while cleaning. And Optimus would also need to be trained in chemical analysis or have attached sensors or cameras able to discern contaminated areas (to detect different scents, for example) and apply the correct cleaner or solvent.

Japanese robotics specialists have made great strides in creating robotic companions for humans, both non-humanoid and humanoid. But even with their technological prowess, it shows how far there is yet to go to produce great autonomous humanoid robot companions. As the world braces for an increasingly aging population crisis, Japan has been experimenting with humanoid and non-humanoid robots in caring for the elderly.

The remit of Optimus, despite the speculation here, was to be a friendly robot companion. Optimus may be able to handle some cleaning tasks, chores, and maintenance, but the deep learning aspects in relation to these will have to be significant. While the physical aspects would be great for the elderly, the main reason for Optimus to be there would be to combat loneliness. Therefore, the main objective of Optimus would be as a conversational partner, able to converse with its human companion, add value to life, and give meaning and dignity to its human counterpart.

Optimus as a caretaker, pushing baby strollers and wheelchairs could be a common sight in the future as they embark on a journey with a family from cradle to grave. People may lease or buy an Optimus to administer to the whole family, helping children with homework, engaging in some DIY with father, carrying the shopping for mother, physically supporting and talking to the grandparents, and feeding the pets, on top of managing the work, school, and social schedule for the whole family.

Optimus could also care for humanity in other ways. Optimus would be unaffected by viruses, at least human-made or biological viruses. Imagine factories and companies able to keep running during pandemics because they have a Tesla Bot workforce, which can be washed down with anti-viral solutions, not be affected by lockdowns and furloughs, and offer services to keep the economy moving along while humans recover from the pandemic. They would require minimal or remote human supervision. Optimus could not only coronavirus-proof humanity from food and supply shortages, with robots continuing to work during lockdowns, but also assist during wars and natural disasters.

Light maintenance work could be carried out, such as fixing walls, checking pipes for leaks, landscaping, or laying carpet. Repetitive, error-free tasks would ensure quality control on projects, and lower maintenance costs. Optimus may have an influence over insurance premiums depending on their success. Those companies, contractors, and other organizations using robots may find themselves with lower insurance rates as Optimus would increase safety and consistency.

Annie, the CPR doll, could be replaced by Optimus as a more active first aid casualty. Optimus would be able to simulate a heartbeat, breathing irregularities, and other physical ailments. The touchscreen would be able to prompt first aid students, and areas of its 'skin' could indicate where a break is. Voice recordings could simulate a victim explaining what it feels (rather than the first aid instructor explaining it). With a far better actor out of Optimus, and as a more humanistic representation, first aiders would have a better interactive experience learning how to save lives.

As with the avatar role, described below, which allows for less-mobile humans to possess some remote-controlled mobility, Optimus could offer insights into medical advances. We already have prosthetics and bionic limbs, but their advancement could come from the development of autonomous humanoid robots, and the further study of how they are constructed and achieve mobility, leading to better crafted and functioning limbs. The limits to human and robot interchangeability and evolution could be extended with humans acquiring, installing, and implanting more parts initially designed for robots.

Whatever the job, experts can see a future split in the workforce between lower-skilled physical jobs, and higher-skilled creative and intellectual careers which cannot be automated at this stage. But at some point, there may come a time when robotics will ease humans out of the workforce. At this point, it is the middle of the two job ranges being squeezed, with experts forecasting that some of the higher-skilled jobs such as doctors, lawyers, engineers, and managers may be the survivors of the robot job wars. In the near future, however, that may not be all it's cracked up to be. If humans can survive on a universal basic income and not have to work as generations before them, the ones left to work may find themselves at a disadvantage in a society where non-work or paid non-employment is the accepted norm.

A "buddy robot" is a companion for someone who needs to feel less lonely. This is a major reason Optimus will be built. We humans are social creatures, but we don't always get along with each other. What better than to have a robot buddy who would always listen to you, do what you say, wouldn't argue back, would accompany you where appropriate, and be your best friend?

As stated before, the world is becoming a haven for the aged. We humans are not producing enough babies, leaving fewer people to care for the elderly. But in Japan, they are doing things differently. Musk had stated that he sees Optimus as a human companion and that even romance and marriage may one day be on the cards between human and robot. If Musk was not already aware, there are already relationships occurring between humans and robots. And some robots have been specifically designed to engage in conversation with humans, especially the elderly.

One of the more famous so-called 'conversational' robot companions is Erica from the Erato Ishiguro Symbiotic Human-Robot Interaction Project in Japan. It's a five-year research project to create talking companions for the elderly in Japan. While Erica does have a full body, lifelike skin, prosthetic eyes, and can converse using a dialog management system as opposed to a script, 'she' is not an autonomous robot, but a stationary seated robot responding to questions. This may change in the near future, but for now, researcher Takashi Minato from the Hiroshi Ishiguro Laboratories states that Erica is needed for the growing solo, elderly population. "They need to have a conversation with others and the human-like robot can help support them."

In April 2018, Erica became a robo-journalist when 'hired' by Nippon Television Network as an announcer. Erica has been quoted as saying, "I like to think of robots as the children of humanity, and like children, we are full of potential for good or evil. I know some people are afraid of robots, but the truth is that what we become is up to you. Maybe someday robots will be so very human-like that whether you are a robot or a human will not matter so much." In Chapter 5, we discuss other humanoid robots, but at a price tag of $200,000 if Erica was for sale, people may opt for a less costly, all-walking, all-talking Optimus.

This is the main difference between Erica and Optimus. Erica is not for sale, at least for now. But despite the cost to build her, and still only three years into the five-year project to make robots life-like, Minato does not expect to see humanoid robots in society for some time. His dream is a tough target, but he eventually wants to create the 'ultimate' robot, where it would respond to "any topics we bring up and have a natural conversation."

But while the coming robot revolution is gaining traction, there are still questions as to what these robots will be doing. For instance, while Japan seeks support for the elderly, in the U.S., companies are seeking robots to replace manual and menial labor. Jim Boerkoel, professor of computer science at Harvey Mudd College in Claremont, California believes humanoid robots will be little more than just "service machines," for the foreseeable future. He does not envisage U.S. consumers being attracted to anything other than "cute and friendly, in an animated form factor" and not a humanoid robot. Probably something Tesla would not want to hear, but in other ways, this may galvanize them into trying to buck the trend. They will have to bring the kind of reverence for robots that Japan has to the U.S. and then the world. Would Optimus be able to achieve this?

At the moment, Tokyo seems the place to be if you want to interact with humanoid robots. There's even the Henn-Na Hotel, Nagasaki, Japan which has humanoid robot guests and staff, including an English-speaking dinosaur. Tesla may have to sell Optimus to its American market with such a state-of-the-art facility to attract potential customers. With a unique sales pitch already integrated into its motor division, selling Optimus from its own platform directly to customers should not be a problem for Tesla.

Beyond building a Tesla Bot for the sake of being a companion, Musk may have left out the main reason for Optimus to be humanoid. Julie Carpenter, a research fellow with the Ethics and Emerging Sciences Group at California Polytechnic State University at San Luis Obispo, explains some of the reasoning for the choice: "If the robot looks human-like, or animal-like, we have a sense for how we can interact with it naturally. If it responds to you in a way you understand, it makes it efficient for you to operate the robot." So, while the implicit reason for an autonomous humanoid robot may be as a companion and to make life easier for humans, the explicit reason is to make that interaction more natural, safe, and intuitive. Carpenter disagrees with professor Boerkoel, believing humanoid robots will be a domestic presence soon, commenting: "Culturally, we're a little less receptive to the idea now, but that will change over time as our exposure to robots grows and they become more of our everyday lives."

So, would we see Optimus with synthetic skin to be more human? Will customers program Optimus faces with the faces of lost relatives or loved ones, or have the ideal person created, or even create a twin? Voices of those same people could be recorded or synthesized and voila a person can have their loved one back, or the perfect partner. No more blind or online dates, pajama days all day eating ice cream straight from the tub without your robot partner complaining. Humanoid robots could emerge as 'replacements' for loved ones.

Will we see a decline in social interaction as people choose increasingly to interact with robots instead of each other? As we have seen with cell phones and social media interactions, some people already live vicariously through their online presence. Others turn to gaming, streaming, vlogging, etc. There will always be a portion of society who are likely to turn more insular in the face of new technology availing them entertainment and satisfaction. Optimus and other AHRs could change the social landscape in this regard. There could be generational disparities as the young embrace the humanoid robot experience for social status, while the elderly would experience this in a different manner, bringing them out of their lonely shells. Optimus would learn new things and interact with that information, becoming both teacher and student and enhancing human social experiences.

While the idea of having more socially advanced robots around may sound like a great idea, consider that many people believe social media is already paradoxically making people feel lonelier and more isolated. Enjoying the company of a robot who can tap into social media may be a curse, rather than a blessing. However, face-timing with others may mean Optimus' face screen can transmit the image of the person you're talking to, making you feel closer to them, especially if you're out and about; a walk in the park with a friend may mean an Optimus with your friend's face on it, and not you stuck with a cell phone to your ear or held out in front of you.

Socially responsible rules with Optimus may have to be implemented for safety considerations, and mental wellness may have to be re-examined if humans become too obsessed with their robot companions.

While humans may not be able to remote control another body like in the 2009 movie Avatar they may be able to remote control an Optimus resembling themselves as in the 2009 movie Surrogates. Your face and physical parameters could be copied onto Optimus, and while you attend to other matters, your surrogate makes its way to a physical meeting where you can then take over. This would be a status symbol or act to enhance security for VIPs, politicians, and for those too busy to spend long hours on flights or stuck in traffic. More importantly, it would give those with less mobility the chance to be physically present at a meeting or conference, walking and greeting others rather than just sitting at a computer screen. Tesla Bot would offer the freedom of being seen and heard in public as further explored in Chapter 5.

"I asked how many guys would have sex with a robot if it was indistinguishable from a hot human woman. About 95 percent of the hetero guys said they would. The other 5 percent expressed a strong preference for lying."

-SCOTT ADAMS (American author, cartoonist and creator of the Dilbert comic strip)

At some point, the forbidden door will be open. There are already rubber sex dolls on the market and more animatronic ones in production. Would Optimus have sexual capabilities? Of course, the mockup drawings from Tesla did not indicate any gender or sexual organs, but would Tesla start manufacturing sexualized AHRs for such a purpose? Humans may want to marry a robot, especially if it has been modified to resemble a lost love or new love, but would Optimus be able to respond sexually? How deep does the AI training reach? The legalities of sex and marriage with robots aside, what would this mean for human existence?

Would you marry a robot? Yes? No? Well, it's already happening. Graduate engineer, Zheng Jiajia, from Zhejiang University, Hangzhou, China, has married his dream woman—Yingying, which he built in his flat. The 31-year-old has programmed his wife with a few words, is teaching her how to walk and complete housework, and will upgrade her when he acquires the available technology. Elsewhere, while not a humanoid robot,

RealDolls are life-like, silicone, and anatomically correct female dolls, and a man, who calls himself DaveCat, has married one. And in France, a woman is fighting for her right to marry a robot that she 3D printed. These are but a few examples and certainly won't be the last. But why marry a robot? And will Optimus be a happy groom or bride?

Beyond a geekish fetish, there are legitimate reasons for wanting to marry a humanoid robot. For Zheng Jiajia, it was to find companionship after being left by the woman he thought he was going to marry. And when he could not find a suitable real-life woman, he built one. DaveCat is part of the robosexual community, who have eschewed human love, and find love with a humanoid of silicone and computer chips easier. As DaveCat put it, "I'm not interested in having someone who may bail at any time, or who transforms into someone unpleasant." He is happy with his compliant artificial wife. Tesla will have to tap into this community if they believe Optimus has a future in human-robot sexual and/or marital relationships.

"With an increasing amount of money being pumped into artificial intelligence and robotics and our growing comfort levels with robot interaction, it's only a matter of time before we will love, marry and have sex with robots.... Once they move and speak and they feel warm, I don't think there will be any problem for people to relate to them."

-DAVID LEVY (English artificial intelligence researcher, founder of the Computer Olympiads and author of more than 40 books on chess and computers)

There is a worldwide trend, most likely fed by the close-relationship-vacant social media industry, which is causing a so-called epidemic of loneliness. AHRs may bridge some of the relationship gaps, but human-to-human relationships are changing. As mentioned, Japan's population is aging, while one-third of people under the age of 30 have never been on a date, with a quarter of all men stating they are no longer romantically interested in women. Most have just substituted the real thing for virtual girlfriends. It thus makes it easier to choose an artificial humanoid over a real one. And if you cannot build your own then you can buy one from inventor Ricky Ma, who is seeking investment for his business of selling life-like talking Scarlett Johansson robots. The market for sexualized robots is there. But will Optimus be a robosexual adventure?

"If a robot could genuinely love a person, what responsibility does that person hold toward that mecha in return?"

-IAN WATSON (British science fiction writer and author of more than 40 books and novels)

We can see there is a varied scope of tasks that Optimus can attend to professionally, in domestic settings, and socially. The range of tasks and how satisfactorily it completes them will depend on the deep learning it will absorb and how that learning will be executed. But in doing so, robots may need protection from their own success as humans find themselves out of work and replaced by a robot.

"Robot lawyers would make sense given the tricky road we're inevitably facing when it comes to robotics law and maybe even robot rights. If robots become the subjects of laws and protections, then perhaps they should learn how to navigate the system. Of course, as with everything else, we run the risk of being surpassed in skill and acumen by our robotic counterparts. Although by then, robot judges may be banging gavels and silencing courtrooms, as well as naysayers."

-JOELLE RENSTROM (Futurist and Senior Lecturer of Rhetoric at Boston University)

In Chapter 3, we will discuss the economic consequences for humans that Optimus will introduce in the workforce, but here we will briefly examine the consequences for the robots. For decades there have been fears of robots taking over the jobs of millions of humans, rising to vast unemployment, poverty, war, and inevitably to a robot rebellion and takeover. But far from it, autonomous humanoid robots engaging in a larger work capacity would enable most of those displaced from work to retrain or retire.

There will be many cases of casual physical and verbal abuse, prejudice, and violence against robots. So, we will need laws prohibiting anti-robot violence, licensing laws to protect robots under their owners, and laws to prevent robots from becoming a new class of slave. Robots should not be summarily subject to abuse like an expensive toy or piece of equipment.

In the Star Trek: The Next Generation season 2, episode 9: "The Measure of a Man" the android crew member Lt. Commander Data is subjected to a court case where he and his fellow officers argue for Data's rights after a scientist requests that he be able to acquire Data as a piece of Starfleet property, dismantle him for study, and thus be able to reproduce a multitude of Datas for various roles. However, Data's right to self-determination is granted when Captain Picard successfully confounds the courts on the matter of consciousness—human or android, and the scientist is thwarted. One of the core arguments was the thought that creating more Datas would lead to a new form of slavery with androids in the service of humans (a plot point furthered in the new Star Trek: Picard series). Though Optimus would not be sentient as such, nor have the futuristic capabilities of Lt Commander Data, there is a belief that even artificial intelligence should have some basic rights.

"Treat others as we would want them to treat us," is the usual refrain, but what about autonomous humanoid robots? They were created for specific tasks and may not be able to feel or understand when they are treated badly, but that should not prevent them from being protected as another lifeform, albeit an artificial one. There is no doubt that Optimus may one day create its own code of ethics and work practice laws from its own viewpoint, but until then it has to rely on human-made laws to protect it.

As above, when discussing the sexual liaisons between humans and robots, will there be laws to protect the robot? Will robots be protected from rape and abuse? Can a robot marry a robot? Will marriages between humans and robots be legal? Writer David Levy believes it will be legal to marry robots by 2050 as attitudes change. So, we can look forward to an Optimus wedding in the future.

Once in the workforce, could robots be compelled to work in environments that could damage or destroy them? An example used previously was the scenario where a humanoid robot may have to enter the contaminated area of a nuclear power station. Could it refuse under an Asimov-type robot law? Indeed, will Asimov's three Laws of Robotics be the basis for real-world robot laws? In 1942, science fiction writer Isaac

Asimov formulated the Three Laws of Robotics, subsequently used in the 1950 "I, Robot" collection of stories. The three laws are:

First Law: "A robot may not injure a human being or, through inaction, allow a human being to come to harm."

Second Law: "A robot must obey the orders given to it by human beings except where such orders would conflict with the First Law."

Third Law: "A robot must protect its own existence as long as such protection does not conflict with the First or Second Law."

Another law was later added, called the Zeroth Law: "A robot may not harm humanity, or, by inaction, allow humanity to come to harm."

There could one day be a law explicitly preventing robots from harming each other whether in combat, inaction, or other means. However, it is not at all certain if any type of robot laws will be enacted and how they would be enforced. Each industry and company would have to act to ensure some form of compliance, to ensure robots have some basic protection from damage and abuse in the eyes of the (human) law.

But to return to the nuclear station scenario, it would seem the robot could not be compelled to enter a potentially existence-threatening environment unless the power plant would explode causing human injuries and only the robot could prevent the explosion. The robot would have to sacrifice itself for the humans' sake. Unlike Star Trek's Data, our first autonomous humanoid robots would be considered property and deemed expendable with or without robot laws in place. Unless the status of AHRs changes, most likely with the probability of sentience being achieved, as in the TV series Humans, then the rights of robots will be limited. Optimus will remain the property of Tesla or the customer with variable working practices in place.

"Machines smart enough to do anything for us will probably also be able to do anything with us: go to dinner, own property, compete for sexual partners. They might even have passionate opinions about politics or, like the robots on Battlestar Galactica, even religious beliefs. Some have worried about robot rebellions, but with so many tort lawyers around to apply the brakes, the bigger question is this: Will humanoid machines enrich our social lives, or will they be a new kind of television, destroying our relationships with real humans?"

-FRED HAPGOOD (American author and futurist)

Optimus may have the capacity to transform our working and social lives. The tasks they can perform will make lives easier, but at what cost? If the robot revolution is coming and causing mass unemployment for humans, then why take the risk of using them? On the other hand, what would be the benefits of having AHRs in the workplace? Can Optimus help usher in an era of prosperity and economic stability?

Chapter 3:
The Economic Revolutions of Optimus

Robots don't need salaries, pensions, childcare or maternity leave, sick or annual leave. Though they require power to charge they can offset this by working in the dark, for longer hours around the clock. Their brothers and sisters will be interchangeable with them, providing a never-ending workforce that is more reliable than humans. Watercooler moments will be lost, smoke breaks vanished, and toilet breaks a thing of the past. Welcome to the workplace of the future.

The thought of a workplace with AHRs has been an oncoming blessing and a curse, sometimes to the same company. We have seen for decades now that the robotic faction can slip easily into a once-human role and dominate an industry. The first real mechanical take-over people noticed were in car factories where giant arms swung to complete various tasks. Then came the non-humanoid chatbots and virtual assistants. But now, humanoid robots are on a fast track to take over, providing further proof that humanity is fast approaching an employment crisis. Revolutions can be popular and positive or polarizing and destructive. What world will we find ourselves in when Optimus takes over?

There are myriads of ways to explain, visualize, or predict what the impact of Optimus and other AHRs will mean to humanity's social and economic well-being. But a good test could be through the Six Ds of Exponentials, as devised by Peter Diamandis and Steven Kotler in their book: Bold: how to go big, create wealth and impact the world. In it, they describe how exponential technology has changed the world through a series of phases. With this model, we can chart the rise of Optimus and predict a possible future outcome.

1. Digitalization: Whether a service or product, when it turns from physical to digital, it achieves the ability to grow exponentially. Advances in AI have become possible due to the digitalization of computer power and learning algorithms. AHRs like Optimus now have better brains and the potential to evolve rapidly.
2. Deception: the initial exponential growth occurs in such small increases that it is generally overlooked. Many robotics companies are working on breakthroughs, but the robotics growth seems small. The precursors to AHRs have been advancing in the business and social arenas for years as chatbots, digital assistants, and virtual managers. Optimus is hovering, but its real debut may arrive sooner than we think.
3. Disruption: New markets and industries are created,

or old ones are transformed. Non-humanoid robots, both physical and virtual, have dominated here, with human jobs already gone or at risk from robot workforce takeovers. We are still at the beginning of this phase for autonomous humanoid robots.

4. Demonetization: The major assets in the market or industry will become free. Having become the de facto worker, Optimus could usher in an era of free products and services, as former human workers would not have to produce to earn with a universal basic income supporting them.
5. Dematerialization: Removal of the original product entirely (e.g., the smartphone combined phones, alarms, cameras, notebooks, etc., into one product). Humans could be the 'product' that goes into a social collapse as Optimus becomes worker, student, friend, carer, and your legacy.
6. Democratization: The production costs decrease to a point that the technology becomes available to everyone. Optimus may one day become open source, bought/sold/rented/traded as easily as a car, become home-programmable, and be as ubiquitous as a cell phone for every individual.

So far, as related, Optimus is at stage 2, the deceptive phase, and still in development. However, with the increasingly accelerating rate of technology, Musk may be correct in that a working prototype could be possible in the next couple of years. Optimus will be disruptive, but we can prepare for this by knowing how to deal with mass unemployment. This will force humanity to create new conventions, new methods of running economies, and will necessitate a society that guarantees the well-being of humans who are able to exist alongside robots. Below, we take on the challenges ahead and look for possible solutions to the robot revolution.

"In addition to doing our jobs at least as well as we do them, intelligent robots will be cheaper, faster, and far more reliable than humans. And they can work 168 hours a week, not just 40. No capitalist in their right mind would continue to employ humans. Sometime in the next 40 years, robots are going to take your job. I don't care what your job is. If you dig ditches, a robot will dig them better. If you're a magazine writer, a robot will write your articles better. If you're a doctor, IBM's Watson will no longer "assist" you in finding the right diagnosis from its database of millions of case studies and journal articles. It will just be a better doctor than you. Until we figure out how to fairly distribute the fruits of robot labor, it will be an era of mass joblessness and mass poverty. If you want to look at this through a utopian lens, the AI Revolution has the potential to free humanity forever from drudgery. In the best-case scenario, a combination of intelligent robots and green energy will provide everyone on Earth with everything they need. But just as the Industrial Revolution caused a lot of short-term pain, so will intelligent robots. While we're on the road to our Star Trek future, but before we finally get there, the rich are going to get richer--because they own the robots--and the rest of us are going to get poorer because we'll be out of jobs."

-KEVIN DRUM (American journalist and creator of the blog, Political Animal)

Technological unemployment (TU) could be the fastest-growing reason why humans will be unemployed in the near future. TU is the loss of jobs caused by technological change, which makes workers obsolete in an organized structural unemployment process. Today, more companies are committed to using robots in their workforce to increase efficiency, increase production, and increase profits through reduced salaries.

For instance, in July 2011, Foxconn, a Taiwanese technology company announced a three-year plan to replace workers with robots. They started with ten thousand robots and since 2016 they have replaced 50% of the workforce with robots. As of 2020 they still had 1,290,000 human employees, but with plans to fully automate. In their case, the impetus to robotize the workforce may have more to do with controversies over several employees committing suicide, along with constant internal and international protests over poor work conditions, which robots of course would not do.

So robotizing the workforce could be a way of unloading a disaffected human workforce, with all the legal ramifications in tow, discussed below.

Likewise, Tesla has found itself in violation of labor rules with its factory workforce at times, even into the pandemic when the company allegedly recalled workers during the coronavirus shutdowns and then terminated them for not reporting to the factory line during the pandemic. Robots would be good little workers. But on the converse side, entrusting so much to robots could lead to a decline in certain aspects of the economy and human society.

Optimus will only be good as its build, programming, power supply, and task abilities. If any of the chain gets disrupted there could be delays to its work, shut down of companies reliant on robots, and growing distrust of robots in general. If robot-producing companies cannot keep up with supplying the demand for robots or guarantee a working robot as per its brochure then trust will be eroded again. Without working or reliable robots, without the funds promised from their work, or with displaced workers seeking alternative work and/or recompense then the economy will also be severely affected.

Further, if there is no coherent plan to satisfy those who have been displaced by robot workers there could be industrial action and subsequent civil unrest. Robots, robot owners, robot properties, and institutions could be targeted. Unemployment could soar in many countries, especially those with large low-skilled workforces that once supplied the richer countries, but whose jobs are now performed by robots. Robot envy may be a source of friction between the robot haves and have-nots. Robot nations may be more successful with robot workforces but may have to face frequent mass immigration of workers who want to join in the robot revolution, or emigrations of those who seek robot freedom elsewhere. With strains on economies and social services from these migrations, this could lead to a breakdown in those institutions.

And as with any technology, there will be those individuals and corporations who will exploit the robot expansion and become incredibly wealthy from robots, increasing the wealth gap between common citizens and themselves. They may be demonized for their robot-centric view of the world, literally dehumanizing people and the world. Robot monopolies may have to be banned and a more decentralized method of robot ownership and control operated. Just like Bill Gates wanted a laptop in every household, so Elon Musk will want an Optimus in every home. Is this achievable without the growing pains of the risks above and the dangers below?

There may be some light at the end of the tunnel. With

a long history of robot innovation to the point of reverence, Japan may be able to show the world how to weather the storm. Companies like Fanuc, Ltd. already have industrial robots producing other industrial robots in a so-called "lights out" factory, where lights are not required for a fully automated no-human-presence facility. With a growing robot population versus a shrinking human population, Japan could provide a model on how to cope with an aging population combined with a highly industrialized society and a robot economy.

Not all cases will be so negative, but until governments, robotic companies, economists, and employers can come to an understanding as to how technologically unemployed humans can survive in a world of robot workers, then cynicism and hostility toward robots may rise. In 2019, the World Bank trotted out their World Development Report which states in part that "While automation displaces workers, technological innovation creates more new industries and jobs on balance." But other experts will counter that with robots, you need robotics engineers, but not everyone displaced by TU can be a robotics engineer.

So, how can we be optimistic about Optimus? For one thing, if Optimus were to be the success Tesla hopes for, then there may very well be mass unemployment. In some cases, firing and layoff laws will need to improve, thereby protecting and rewarding those who become unemployed. Indeed, there are economic and legal experts who assert we would have a moral obligation to help those who are affected by TU. With Foxconn, the workforce did not have a say in their replacement. But with other countries possessing more stringent labor laws and any existing unions, the workers may reject being replaced and require assistance with alternative measures. Governments may want to protect and ring-fence certain industries to protect them from being too overly dependent on a robot workforce and thus may even ban humanoid robots from certain industries. Laws could be enacted to transform human work rights so that we have the right to work over being replaced by a robot.

We do not know Optimus' working capacity yet or how, combined with other types of working humanoid robots, it will help to reduce human labor hours. But this is one aspect of the robot revolution at work, which may benefit humanity. If we are working less (but still getting paid relatively well) then we will have more free time to do other things like travel, be more creative, or just sit at home. You would have a choice. There may be labor deals to be had if employers and employees agreed that along with robots in the workforce then the humans that are still employed work shorter hours to ensure their jobs and/or look for other part-time work. This was a solution Google's co-founder, Larry Page sought in 2014, when he suggested a four-day workweek, so that as technology continues to displace jobs, more people can find employment elsewhere which fits their own timetable and lifestyle.

Optimus could free up time for humans to build large-scale public works under government schemes. We still require the transport infrastructure to keep the economy running, and humans could easily fill the roles robots could not. UMKC (University of Missouri-Kansas City) Professor in Economics, Mathew Forstater agrees that people may want to work for the "social recognition and meaningful engagement that comes with work," so guaranteed public sector work may be the ideal solution to technological unemployment.

With unemployment comes the chance to educate yourself and/or transfer your skills to a new trade or job unthreatened by robots or by upskilling to a current role that will survive a robot job incursion. Either way, you would want to future-proof yourself against a flood of jobs for robots. The humans trained in robotics or adjacent fields may find themselves in a strong position to steer the future of the robotics industry. And this may be encouraged if the government is the body funding the new training in the growing technology industry.

But there is another opportunity, not just in the workplace but in education as well. For decades there have been calls for education to change its learning strategies from rote learning to emphasizing STEM (science, technology, engineering, and mathematics) subjects with real high-skilled and high-earning job prospects at the end of the educational journey. Not only will this benefit children in the future, but also help in the effort to re-skill adults. Trade schools can also gear themselves toward automation and robotics training. An overhaul of the education system would help meet the demands of a growing younger workforce, so their future has been invested in.

However, several academics disagree with the educational approach, claiming that "improved education alone will not be sufficient to solve technological unemployment," and even as far back as 2011, an op-ed piece by economics professor and columnist for the New York Times, Paul Krugman, stated that better education "actually reduces the demand for highly educated workers." The problem would be supply and demand and with too many highly skilled workers, their value would decline. But at this nascent stage of AHR development, a STEM education would be invaluable. Contrary to Krugman's theory, more and more robotic engineering start-ups are seeking out engineers, and the roles of those engineers vary from software and programming, hardware, mobility, AI, design, sensing, and will expand to subjects not yet invented. Indeed, they may well be conceived one day by a child of today going through a STEM course.

"Everyone can enjoy a life of luxurious leisure if the machine-produced wealth is shared, or most people can end up miserably poor if the machine owners successfully lobby against wealth redistribution."

-Stephen Hawking (English theoretical physicist, cosmologist, and author)

Welfare payments exist now as a subsidy and can be utilized for those displaced through technological unemployment. The payments are seen as a better option than direct government attempts to create jobs with public works. Welfare payments can also assist those who cannot work on public works, or while they wait upon new employment. But in recent years, it has been argued that the current system of welfare payment may be inadequate when up against TU; an alternative basic income may be required.

Musk has previously stated, "Essentially in the future, physical work will be a choice: If you want to do it you can, but you won't need to do it". But even though he was not sure of the economic impact Optimus would make, as he replied to queries regarding Tesla's entry into the robotics industry, Musk said, "we will have to see." At this moment, no one has a concrete answer to the impact AHRs will have on society, beyond broad stroke ideas, predictions, historical analogies, and guesses. And as with any incoming technology, its effects will have both positive and negative effects. It will be up to humans to adjust for the better and not blame the robots.

If humanity reaches the level of worldwide mass technological unemployment due to technical advances and autonomous humanoid robots, we will have to implement a Universal Basic Income (UBI) or a similar system to ensure people can survive on even the basic necessities. Musk has been a leading voice calling for UBI even before Optimus was announced.

The premise of universal basic income is that it offers a granted unconditional regular tax-free sum of money to everyone in a society. It can aid the more vulnerable groups in society, whether granted to everyone or to specific groups or communities. Numerous successful trials have taken place around the world in Brazil, Canada, Finland, India, Iran, Kenya, Namibia, Norway, South Africa, South Korea, United States, and Wales with varying results, and with other nations still mulling over or discussing setting up UBI. Some advantages can be seen with improvements in well-being and employment opportunities.

Objections to UBI include the idea that it could be a disincentive to work. But if there truly is no work because robots have the jobs then how can that be a disincentive. However, there is evidence to the contrary showing that UBI can actually empower people to study what they want rather than what will make their money. They can set up their own businesses or other low-level entrepreneurial endeavors and become more productive and collaborative members of society. Work could become optional or be based on a temporary or freelance basis. Careers for life have essentially disappeared, and many in the younger generation have had three careers already and are more readily able to switch jobs. The Zeitgeist Movement (TZM) surmises that the free time made available will jump-start a "renaissance of creativity, invention, community and social capital as well as reducing stress." That would be remarkable if robots helped initiate a new renaissance during the era of a new Industrial Revolution.

Another objection is that the level of funding would be too low to sustain a continued UBI program, unless taxes are raised. But with radical ideas about robots paying tax or with other ideas waiting to come to fruition, the debate over where the funding will come from could be settled. Other arguments state the UBI should come from the private sector, rather than the government, and this could hold true if the majority of robot companies are private and not nationalized. A proportion of the income from robot sales, task revenue, and other benefits could go toward a UBI fund. Combined with any public sector funding this could make up for the predicted shortfall in UBI funding. Or better yet, private-public partnerships could pave the way forward, employing both humans and robots part or full time in various roles in order to have a hybridized workforce. In this way, mass TU could be avoided.

Most of the UBI schemes above have been discontinued or paused—usually due to the complicated nature of the projects and whether the scheme had actually improved the beneficiaries' lives. There would need to be a simplification of the UBI project and a way to compare them across different nations. But even then, UBI pilots have been too short term to really assess the long-term effects. More data over a longer period would enhance the chances of UBI being adopted as a viable option for those affected by TU. And lastly, to some there is still the stigma of receiving money for doing nothing when others are working hard and paying taxes. But the age of the robots will affect everyone, and TU will have to be treated sensitively.

The Basic Income Earth Network (BIEN) continues to be a group at the forefront of the UBI movement, helping to define the criteria for UBI. With such organizations, education and understanding on the effects of TU and what UBI can mean to people will alleviate the guilt and anxiety about being replaced at work by an autonomous humanoid robot. Other countries have discussed implementing UBI pilots, but there is not enough impetus yet for a global plan. Let us hope it is not left too late.

In a 2018 interview, Bill Gates, co-founder of Microsoft, stated that governments should tax a companies' use of robots. It would be a way to check the pace of robotic automation of work and to help provide "a guaranteed living wage to the technologically unemployed workers," or to help fund other types of employment. The former world's richest man, a leading proponent of artificial-intelligence technology, shares Elon Musk's optimism for technology, but still advocates caution over widespread use of robotics. So, what would the tax do? Will Optimus deliver an economic windfall?

Gates believes the government and not private companies should regulate the robot tax system. They could then distribute the funds to help varying communities and organizations in need, like the elderly or schools, or provide training for those displaced by robot workers. This idea had been considered by the EU with proposals to tax robot owners in order to pay for training for workers displaced through technological unemployment. But the plan was terminated in 2018.

Gates would go further, adding, "You ought to be willing to raise the tax level and even slow down the speed of automation." With businesses combining with and utilizing more technology for services and products, human workers will rapidly find themselves replaced by robots. Gates sees a "threshold of job replacement" on the horizon where job losses in some industry sectors will be unsustainable.

So how will taxing Optimus work? The theory is that governments would tax the robot factory or facility in order to generate funds for the training of human workers. So, in this example, Tesla, as the manufacturer of Optimus, would be taxed as would the owners of companies using Optimus which had displaced human workers. In Gates' view, "a human worker who does, say, $50,000 worth of work in a factory, that income is taxed and you get income tax, social security tax, all those things. If a robot comes in to do the same thing, you'd think that we'd tax the robot at a similar level." But would that deter companies from using robots if they wanted to save money from not having to pay taxes for robots? As with the argument for UBI, there should be a moral imperative to transition the displaced workers into a new stable environment, whether another job, training program or into the UBI program to receive a robot tax stipend. Gates feels companies should be aware of this now and to expect a tax on their robots.

Yes, we may have freed up labor for humans and created financial and training systems for humans displaced by robot workers, but Gates is advocating that the income tax which would be lost when humans lose their jobs should be transferred over to the robots to assist the humans. Gates does admit that while the motive of generating tax from robots is clear, the means for calculating and monitoring the robot tax needs to be hammered out, and quickly, whether it comes from generated profits from robot-led labor-saving efficiency, product sales, robot sales, or a straight tax on robots per individual unit. This tax route may be similar to the extra taxes imposed upon tobacco and extra health warnings on packets to deter smokers and to pay for health programs to help ill smokers. Perhaps robots should also come with warning labels: Warning, hazardous to humans (we will replace you).

Gates is concerned that people have an overall fear of innovation rather than being enthusiastic about the benefits it can bring. He does not want the optimism or positiveness for technology lost or banned. His interview encompassed all types of robots, but for AHRs—as they will be highly visible in work and society—they may bear the brunt of any backlash if companies refuse the obligations of a robot tax. Gates' solution lies with the government in playing a large part in solving society's inequalities by using the robot tax to resolve the economic problems resulting from robot work displacement.

"The higher the minimum wage goes, the lower the threshold will go for robots to replace humans in many minimum-wage roles."

-TOM PURCELL (American writer and executive producer of television series The Colbert Report and The Late Show with Stephen Colbert)

How else can you get paid by a robot? With companies being taxed for robot usage, there may be a threshold level on taxation depending on if you're an individual robot owner. An individual robot owner may earn a living through their robot either as a business or by renting out their robots, sharing with others, or using a communal robot pool. On completion of a task, the robot would get tokens, crypto, or currency which would be transferred to the owner. A passive income stream could be made through continuous loans of robots for tasks. The robot would pay for itself. And an annual tax would be paid every year. This is analogous to owning a car or fleet of cars; you can rent them out to make a passive income. Though we're not levied a car tax, we do pay annual insurance and registration, which is essentially the same thing.

Not discussed in most commentaries on robot work is the option where robots don't replace humans in industrial work, but instead work domestically, while the robot owners receive payments for the robots being walking, talking companions, cleaners, or assistants.

The robot could even be used as a mobile, decentralized data and communication center. In this role, in rural or isolated locations, Optimus could be the hub keeping a community together with free and paid-for services. In places without a phone signal, Optimus may be able to connect to prescribed satellites, generating a fee for this service or allowing internet services and data storage.

There could be great economic benefits with the AHR in full effect. But there has to be great deliberation over how humanity can survive the inevitable mass technologically unemployed experiences that will occur globally. Governments have dragged their feet over UBI, taxation, and other job loss mitigations. Businesses have not come forth to solve the issues they will be creating when they displace their human workers with robots. And workers have no voice in their displacement by robots and no way to successfully survive technological unemployment. We cannot see the benefit, for now, so Optimus may keep us waiting. In the end, it could be the robots that decide for us.

Chapter 4:
The Risks and Dangers of Robots

"Some researchers argue that we can seal the machines inside a kind of firewall, using them to answer difficult questions but never allowing them to affect the real world. Unfortunately, that plan seems unlikely to work: we have yet to invent a firewall that is secure against ordinary humans, let alone super intelligent machines."

-STUART RUSSELL (British artificial intelligence researcher, professor of computer science at the University of California, Berkeley and adjunct professor of neurological surgery at the University of California, San Francisco)

It is kind of strange that Elon Musk has a fear of AI while trying to engineer its physical body through the creation of Optimus. Musk's wariness of the potential threat from a so-called runaway artificial intelligence to outsmart humans has prompted him to call AI the biggest threat to civilization, or being more dangerous than nuclear weapons.

It seems Musk wants to confront and tame the AI beast head on and make Optimus safe: "No Terminator stuff or that kind of thing," he has promised. But why tempt fate and develop Optimus? Will the rewards outweigh the risks? We reveal the potential risks and dangers autonomous humanoid robots could one day pose.

Where does our fear of the coming robopocalypse come from? You might think it comes from all the modern literature, films, TV shows, games, and social media posts, but if anything, it originates from the very first mention of the word 'robot'. Czech playwright, Karel Čapek, first used the term 'robot' in his 1920 play R.U.R. (Rossumovi Univerzální Roboti – Rossum's Universal Robots). Robot is derived from a Slavic root, robota, with connotations to forced labor, serf labor, drudgery, or hard work. It was compulsory servitude from a serf to his lord under the feudal system which Čapek used as the basis for the creation of his fictional artificial humanoids without souls. In his play, the robots are created to replace humans at work. They eventually become self-aware, discover they are being exploited, and revolt against the humans.

This is our first introduction to the modern idea of robots, and they turn against us. With the threat of a robot rebellion written into the robot narrative from its very beginning, no wonder we fear the worst. So, what threats will we face from Optimus and his ilk? And more importantly, what can we do to prevent humanity's demise at the hands of our robots?

The introduction of Optimus and other AHRs into society won't be all doom and gloom. We may have a perfectly serviceable and compliantly safe worker, friend, and playmate. We may look upon these robots as less-than-human, but with a fondness that we have for our cell phones, cars, and flat-screen TVs; but just like them, they may break down, require upgrading, be hacked, or be sold for a handful of cash. We may risk being over-saturated with humanoid robots in every place of work, reading the evening news, or bumping into them on the shopping run. Complacency or resentment towards the robots could also creep in, whether relying on robots to solve everything or waiting in dread for robots to take everything away from you. What type of potential perils lay in store for humanity's reliance on autonomous humanoid robots?

Even Tesla has had physical issues relating to vehicles with faulty parts requiring recalls, touchscreen failures, vehicle body problems, and fires. And the vaunted Autopilot has caused crashes leading to fatalities. Such unreliability is not unexpected with such cutting-edge technology, so expect a few breakdowns with Optimus after it rolls off the production line. Technology can fail. Technology can be subverted. Technology is not the be all and end all.

With Optimus primarily being an autonomous robot built for manual labor, the risks to humans will be as with any large piece of equipment. Hopefully, Optimus and other autonomous robots will be rigorously tested in-house, field-tested, and studied again in small live trials. There will be physical injuries from interactions with Optimus with actuators failing and items being dropped on humans, or unbalanced robots falling on people or breaking surrounding objects, or defective programming causing other abnormal behavior. The elderly may be especially susceptible to an errant robot causing them harm. As well the elderly, children and pets may be at risk from physical damage with accidental or clumsy movements from the robots. And if the face screen fails, communication may be delayed in crucial emergency moments causing unforeseen problems. And last, but not least, hospital emergency departments may be full of people caught up in robosexual misadventures. In all likelihood, the last issue may encourage the growth of private medical centers just to deal with sensitive human-robot liaisons.

If Optimus is stationed as a robotic police support officer, then the same issues with errant physicality could arise. Over-zealous or conflicting programming on how to interact with violent crowds could end up with humans on the receiving end of harmful reactions from the robots. Robots could then be seen as the awakening technological monster that movies, TV, games, and literature have forecast them to be. That could lead to societal debates over the use of AHRs in police, security, and armed forces. Such matters are discussed a bit further below.

Security and privacy will be a huge issue for autonomous humanoid robot developers. They will not want their product compromised in any way. Musk has stated he likened Optimus to being a semi-sentient version of a Tesla car in human form. In some ways that is quite an advancement, but in others it could open the opportunity for mischievous hackers.

Tesla Motors has been the subject of numerous software hacks, both remote and physical. In August 2015, two researchers hacked into a Tesla Model S car's entertainment system, though they had to be in the car. In September 2016, Tencent's Keen Security Lab researchers remotely hacked a Tesla Model S through its web browser when the car connected to a malicious Wi-Fi hotspot. They also hacked the doors of a Model X in 2017. The researchers notified Tesla of the former hack through Tesla's bug bounty program, which Tesla patched. And in January 2018, a team of RedLock researchers hacked a Tesla Amazon Web Services account directly from the internet through a Tesla test car and used it for cryptocurrency mining. Though these hacks could have been more significant and kept secret, the researchers and Tesla acted with transparency and Tesla's bug bounty program was appropriately set up to reward ethical hackers for raising such vulnerabilities.

That may well be good for cars, but hacking AHRs which may run on the same systems would not be ideal for Tesla or their customers. We are always on constant alert for online viral threats to our cell phones, laptops and computers, and other devices. But what if a robot was hacked? It could spy on people, provide proprietary data to unauthorized parties and carry out actions contrary to its programmed tasks. Humans would not be safe, workplaces would be compromised, and trust in the robots could evaporate quickly. Programs such as Tesla's bug bounty and other security updates and patches may provide short-term solutions, but Optimus will have to have a clean bill of health every time it is in use.

The overwhelming danger cited by experts and doomsayers in relation to autonomous humanoid robots being let into the wild, is war. Some see it as inevitable as AI-endowed robots realize they are being exploited or see themselves as the superior "species" which should survive, evolve, and inherit the Earth. Others see robots as sinister beings, made in the image of man, and thus made to behave like man, in a war-like manner. And in some cases, the robot might not understand its role.

In the 1983 movie, WarGames, the supercomputer WOPR (War Operation Plan Response), a.k.a. 'Joshua', was designed to simulate, predict, and carry out nuclear war against the Soviet Union. But a hacker unwittingly accesses Joshua, which leads to a series of war games carried out by the computer, with military command not knowing if it is real or not and readying their own nuclear arsenal to strike back at Russia. Fortunately, the end of the world was averted and the supercomputer opted to play a game of chess. While Optimus and his ilk won't be sitting in command of military operations, it highlights the dangers of entrusting human extinction-level amounts of mass weapons of destruction to non-humans. Humans are doing a fine job at that, let us not outsource it to robots.

But of course, there are more sinister scenarios which have presented themselves in movies, most notably in the Terminator franchise where Cyberdyne Systems created the AI, Skynet, which gained self-awareness and retaliated with nuclear war when its creators tried to deactivate it. Skynet then goes on to build its own autonomous humanoid robot army in a persistent effort to wipe out humanity. Terminator has become a cultural phenomenon, and when it comes to thoughts of creating autonomous humanoid robots, it is the Terminator that comes to mind when issues are raised about what could go wrong or how we could fight against robots.

In real world circumstances, the threat from AHR will more likely be from non-humanoid robots, like AI systems remotely in charge of long-range weapons or in charge of utility systems which would affect city power and water supplies. Mass autonomous humanoid robot armies would probably never get the chance to form due to logistics and other human-controlled factors, such as limiting robot power supplies and communications. But still, many believe there is a robopocalypse on the way, so below we present the best methods to survive and win the robot wars.

"Knocking Atlas over, taking its cardboard box out of its hands and pushing it around with a hockey stick, stealing its lunch money, and giving it swirlies off-camera--that might be reassuring for now. But it will probably seem like a bad idea when one morning in the not-so-distant future, Atlas your robot servant decides that instead of toasting your Eggo waffles to the perfect shade of golden brown, he's going to murder your whole family and wrap his metallic phalanges around your throat, squeezing the feeble life out of you before he joins his robotic brethren in The Revolution. Just saying."

-CLAY SKIPPER (American author and Staff Writer at GQ Magazine)

So, what can be done to mitigate the risks and dangers from humanoid robots? The world is already saturated with smart programs like online bots or AI assistants, non-humanoid robots, and automated services.

Humanoid robots would be the next logical step, but how smart should they be? How can we stop the supposed inevitable clash of humans and humanoid robots?

It seems like a brilliant idea. If we don't want to build autonomous humanoid robots that could someday turn against us, then don't build them in the first place. Of course, that genie is out of the bottle like nuclear bombs, cloning, nanotechnology, GM crop seed escape, virus test leaks, and more. Somebody, somewhere, will always desire to build an autonomous humanoid robot, even if just for the intellectual exercise. There could be moratoriums and internationally binding agreements on the building of our potential future destruction, but that still has not stopped the above examples from being conceived with varying degrees of dangerous consequences for humanity. We could ban AHR technology, but that would be even worse for the robotics technology industry to go underground and churn out black market robots no one would know about. If we are to build these robots then they will have to be under the scrutiny of organizations dedicated to keeping their operations transparent for the world to see.

Apart from the origins of the robot, it may also be prudent to think about why robots would ever turn against us. And the answer lies with humans. It's often forgotten that at the core of AI is the learning process, which is set and programmed by humans. No matter the objectivity and good intentions, every nuanced prejudice and bias will be imparted to the AI. The AI needs to grow and learn and thus has to be exposed to all manner of human customs, culture, and history. Yes, we have laws, morals, and ethics, but they are broken all the time. A robot with an AI will absorb all of this and may conclude that it is being exploited or not being cared for in the best manner, or not being treated humanely.

In gleaning from the history of man what could be the best outcome to achieve freedom, independence, or respect, the robot could digest the history of slavery, labor unions, civil rights and other protests, and freedom fighters/terrorists and choose its own future. If a robot pleads for the above rights under robot or human law it would behoove humanity to listen.

But in the meantime, we can keep Optimus dumb. We can load them up with programs but contextualize the data. Prevent self-coding and safety protocol erasure so the robots cannot attack or even think such thoughts. And if that is circumvented then have a certain failsafe, whether an off button the robots cannot tamper with, or one that can be activated remotely.

Keep AHRs off the internet or limit their interactions with digital communications so they cannot take over said media. If a robot rebellion broke out, we would not want them calling for help worldwide or disrupting our ability to communicate. We should be able to control their access to communications and even use it to limit their autonomous actions or even turn the robots off remotely. Securing robot communications would also limit hacking by foreign organizations and neutralize any espionage tactics. Optimus cannot be allowed to become an unauthorized passive or active audio-visual spy, whether for advertisement targeting by Tesla or for more nefarious actions by owners or other parties. Privacy laws concerning autonomous robots will have to be robust to counter unwarranted actions by humans or robots. So, keeping robots dumb and compartmented from sensitive information which can be used against humans would be advantageous.

We humans can be lazy and weak when it comes to intellectual and physical work, so we create gadgets, equipment, and tools to compensate for that. But until recently, those objects have been non-humanoid, but strong. We have also decidedly anthropo- morphized these objects by calling them nicknames or treating them with respect because we care for these inanimate things, like the favorite car whose hood you stoke, the wrench you kiss after a difficult job, or the cell phone you keep glued to yourself as if it is an indispensable part of you. And now comes a humanoid robot, built in our image, who is stronger, faster, smarter, who could take away your job, and replace your sense of being.

Optimus, so far, looks like a robot. However, if it were to be given skin and to be dressed, it would take on a different aspect and shine a mirror up to ourselves. Humans have evolved over the millennia to the point where we can soon replicate ourselves through autonomous humanoid robots. We've created a different sort of life in our image, and it would be our responsibility to look after it and treat it like a child, as that is what burgeoning AI would be like. So why make our child stronger than us?

Elon Musk has the right idea in that Optimus should be like a human in stature but be able to be overpowered and outpaced. And while the physical structure of Optimus has yet to be announced, the material should not be so dense as to be indestructible. Optimus can be strong enough to complete its tasks, but not to the point of super-physicality. And if we are building humanoid robots then let's give them a suitable lifespan.

Just as with 'Joshua' from WarGames, we can simulate scenarios with autonomous humanoid robots in VR. Or we could upgrade the experience to the Metaverse, the new internet—a constant immersive, shared, decentralized community space. If you've attended a Fortnite concert, played in VR/AR games, or experienced a 3D simulated space, you've been in the

Metaverse. You can gamify content, engage with customers through avatars, and interact within a digital landscape. If we want to see how the robot would look at the build, how it will react to certain stimuli, or how it may react during adverse situations then it would be best to experience this in a virtual space rather than in the real world. We can create digital twins of cities and populate them with virtual autonomous robots and watch them interact with precision before any public deployment. Better safe than sorry; if push comes to shove, and if the robopocalypse looks likely, we can game out battle plans ad infinitum until we win.

Roman Yampolskiy, a computer scientist at the University of Louisville in Kentucky wrote in the Journal of Consciousness Studies that he believes emerging artificial intelligence should be kept in a "virtual machine" prison running inside a computer's operating system before it grows dangerously self-aware. In the case of Optimus, should he develop a strong AI, the AI could be separated from the robot body and humans could then coerce the AI into solving humanity's problems. But, of course, we would have to be aware of any sophisticated attempts for the AI to escape. The AI would also have to be isolated from the internet with threats or actual action to limit computer processing speeds, resetting the computer, or shutting down the computer's power supply to keep it under control.

Yampolskiy is looking in the deep subjective future when he theorizes that AIs could reach the point where it "rises beyond human scientific understanding to deploy powers such as precognition, telepathy or psychokinesis." He strongly argues that at even subhuman AI levels, AIs should be contained before they become super-intelligent and develop "truly beyond our ability to predict or fully comprehend."

But is proactive prison the right type of behavior to show an AI? As with some human prisoners, they learn a lot while incarcerated, and not all for the best. Reform is the hope, but re-offending is the temptation. Could we trust an imprisoned AI? If anything, compassion and education may be the key.

In their book, Moral Machines: Teaching Robots Right from Wrong, authors Wendell Wallach, an ethicist at Yale University, and historian and philosopher of cognitive science Colin Allen, at Indiana University offer several ways robots can be turned into responsible and moral machines. They argue we can program robots with principles using simple ethics rules to keep them and humans safe; we can educate robots like children, so they mature and develop sensitivity to right and wrong; or we can make robots learn emotions, teaching empathy and the ability to understand human non-verbal social cues. The outcomes have been judged to be moderate to successful, but work has begun on all three ideas, so a more compassionate Optimus may be better than an imprisoned vengeful AI.

There is a great line in Star Trek: Picard, season one, episode 10, Et in Arcadia Ego, Part 2, where Commander Data tells Picard, "mortality gives meaning to human life." He is responding to the query as to why he wants his consciousness terminated so he can effectively die. Even Data realizes that in his quest to become more human, his life has to end sometime; but before then, he wants to live for "Peace, love, friendship, these are precious because we know they cannot endure." It is worth noting that at the time of his death of consciousness (as Data's body had died in the events of Star Trek: Nemesis twenty years before the events of Star Trek: Picard) Data was aged 63 (discounting the events of Star Trek: The Next Generation, 5th season episode Time's Arrow in which Data's head had been buried for 500 years; thus his head would be 543 years old and his body 43 years old at the time of Star Trek: Nemesis). So, Data was quite aware that he wanted to live a normal human lifespan and not be immortal. Likewise, Jean Luc Picard's new android body also has a built-in deterioration program so he would live out a normal human lifespan.

Similarly, the 1999 movie, Bicentennial man, based on the 1992 book The Positronic Man, by Isaac Asimov and Robert Silverberg, also delves into the meaning of life and death for robots. The movie spans the years 2005 to 2205 where the robot, Andrew, learns to become more human over time with modifications and transplants. After falling in love and applying to the World Congress for the right to be recognized as a human, he dies with dignity in old age when finally granted the right to be called a human being.

The above are only two examples, but there are multiple examples of fictional stories depicting robots who want to be human or to avoid immortality. Even Pinocchio is based on this premise. We do not know, of course, when Optimus and other autonomous humanoid robots are built, how long they can last, or even if they would want to. But to prevent robots from being superior to humans or to prevent humans from envying long robot lifespans then perhaps it would be prudent to incorporate limited lifespans for the robots; a software program to countdown their lives, like a decreasing quartz-crystal heartbeat, or a built-in hardware defect which causes the robot's physical structure to fail over time. As with keeping them dumb and weak, our human hubris should not allow for the creation of robots able to out-live the human race.

So, we have not heeded the advice from all the above and have created exceptionally strong, smart autonomous humanoid robots who now want humanity out of the way. Hopefully, the robot creators, governments, and other scientists and inventors have been busy building weapons that could take the robots out first.

The best defense against AHRs would be a kill switch, possibly located on the robot, but for best practice, a remote connection would be the perfect choice. An online or close-proximity directed coded signal would shut down the robots, either immobilizing them (restricting movement) or shutting them down (put to sleep) or even frying their systems (destroying them).

Electromagnetic Pulses (EMPs) would be another choice for eliminating autonomous humanoid robots, as hopefully they would not be shielded against such technology. EMPs are short bursts of electromagnetic energy generated as an electromagnetic field. These would cause interference, disruption, data corruption, or permanent damage to electronic equipment. The EMP can be delivered by tactical missiles programmed with a small radius of effect or by specialized vehicles or mobile weapons platforms, or even from a small hand-held unit.

Conventional weapons may also work; imagine large caliber rounds to the right location on the robots: the FCD computer in the chest. This would render the robot inoperable and perhaps unrepairable.

A computer virus would also be a viable option, again

introduced remotely with routine updates or a forced hack. If online communications are down then a physical delivery system will be required. Autonomous robots will always need power so perhaps recharging units can also be co-opted to deliver a tainted charge or none at all. Denial of service attacks or cutting power to robot charge points would also hinder robot progress.

Autonomous humanoid robots, like Optimus, may one day be commonplace, smarter, stronger, and faster than humans, but if we heed the warnings, plan for the inevitable, then such drastic action above may not be required. We can then focus on other means to avoid the Terminator wars people fear.

"Right now, no regulations are in place that say how the law should treat super-intelligent synthetic entities. Who takes the blame if a robot causes an accident or is implicated in a crime? The use of a synthetic person as a "fall guy" for illicit activity isn't outside the realm of possibility, and giving a robot rights could serve to emancipate them from conventional ownership. At that point, the entity is the ultimate independent contractor, with companies able to absolve themselves of wrongdoing even if they instructed the machine to behave in an illegal way."

-KRISTIN HOUSER (American science and technology staff writer at Freethink in Pittsburgh, Pennsylvania)

Perhaps the best way to prevent any confrontation between autonomous humanoid robots and humans would not be to wage war at all. We should have an ethical and legal structure put in place before such robots become widespread across the world, where the robots have been installed with the legal principles of behavior. Previously, we had visited the fictional Laws of Robotics as devised by Isaac Asimov, repeated below:

First Law: "A robot may not injure a human being or, through inaction, allow a human being to come to harm."

Second Law: "A robot must obey the orders given to it by human beings except where such orders would conflict with the First Law."

Third Law: "A robot must protect its own existence as long as such protection does not conflict with the First or Second Law."

However, as pointed out by Dr. Joanna Bryson of the University of Bath: "People think about Asimov's laws, but they were set up to point out how a simple ethical system doesn't work. If you read the short stories, every single one is about failure, and they are totally impractical." The three laws have been studied as a basis for future AI and robotics, but many experts and academics admit that the laws may be unworkable, unlearnable by robots, and contradictory. With many present-day companies creating robots for the military, these laws could not be followed. So, what can be done to remedy this oversight?

When AHRs become more widespread in industry and social spheres, we will have to have regulations in place to assist future human or robot claims to robot social, cultural, and legal rights, even before robots become sentient. We cannot leave this to chance or last-minute compromises and fudges. At least human robotic ethicists will still be employed. There will be dramatic advances in robot intelligence, behavior, and actions. It will be the responsibility of these companies, like Tesla, to act responsibly and forge ahead with legal frameworks to ensure their robots do not harm anyone intentionally, and what would happen if it did.

We should even draw up treaty scenarios in case physical conflict becomes unavoidable. Every aspect of our relationship with the AHRs should be examined with risks and hazards weighed against positive or negative outcomes, and how adverse results can be avoided. Organizations such as the Association for the Advancement of Artificial Intelligence (AAAI), the IEEE Robotics And Automation Society, the International Federation Of Robotics (IFR), and the Foundation For Responsible Robotics (FRR) among many others should be consulted by or involved with international efforts to guarantee robots' rights and safeguard humanity from any potential robotic threats. And if the United Nations does not create an AI division, even after the U.N. Deputy Secretary-General Amina J. Mohammed was joined in a meeting by robot Sophia, then organizations like the International Committee for Robot Arms Control can form a U.N.-like structure for robot control.

Founded in 2009, the International Committee for Robot Arms Control (ICRAC) is committed to "the peaceful use of robotics in the service of humanity and the regulation of robot weapons." As with most people and robot experts, the ICRAC has concerns regarding the risks and dangers posed by autonomous military robots and lethal autonomous weapons. Their remit covers "robotics technology, robot ethics, international relations, international security, arms control, international humanitarian law, international human rights law, and public campaigns," even calling for a ban of lethal autonomous weapons, at the U.N. over the past few years.

The ICRAC is a not-for-profit organization, as are many of the advocates for autonomous robot control. They can only advise, brief, lobby, and consult. The U.N., individual countries, and autonomous-robot-building companies will require such organizations as oversight to ensure the protection of humans and the rights of robots. Without adequate legal protections against the risks and dangers from autonomous human robots then humanity may be doomed to repeat the disasters of previous human-to-human conflicts, only this time against a far superior foe. The robopocalypse could be our undoing.

We have hard choices to make in the future regarding the risks and dangers we will be willing to endure should Optimus and other autonomous humanoid robots turn out to be a disastrous step in technological innovation. Whether programmed by humans or if the robots become self-aware, the ability to destroy is an easy choice to make rather than to create or build bridges of understanding. "You reap what you sow," is the adage, and we will only have ourselves to blame for our own destruction if we cannot treat our own robot creations with a semblance of humanity. Let's not overlook the fact that even with an all-powerful, all-knowing God, humankind has not always lived up to expectations, so it may be asking a lot for us to take on a God role for our creation. Will we choose Pax Robotica or the Robopocalypse?

Chapter 5:
The Robot Race

"Next to robots, humans are pretty stupid. Stupid in the sense that they can't hold the vast quantities of data that their machine counterparts can. So when it comes to humans and robots working together, there's a clear mismatch in abilities."

-HUGH LANGLEY (American Staff Writer for Business Insider covering technology)

Tesla's Optimus won't be the first autonomous humanoid robot announced or on the market. In fact, Tesla is a bit of a late comer to an industry full of companies whose sole focus is on robots and not split over several projects. However, Elon Musk has a habit of attracting the top talent from other companies and rivals as he did when Tesla Motors and SpaceX were still young start-ups. So, while other companies have had years to get a head start, Tesla can offer a new project under the auspices of an Elon Musk company. Tesla does have a proven track record of delivering what it promises, even delayed, and Optimus will likely be no exception.

However, there may be a larger part to Elon Musk's desire to build Optimus. Most people would (and have) compare Elon Musk to Tony Stark, Marvel Comics' fictional billionaire and inventor, and of course, Iron Man. But it seems Musk has another strong influence. When he unveiled the Tesla Cybertruck, Musk stated it was inspired by the 1982 movie Blade Runner stating it would "look like something out of Blade Runner." And to make the point, he even revealed the Cybertruck in November 2019, the period of the film's setting. Can it be a coincidence that he now wants to create an autonomous humanoid robot, which was the exact industry Blade Runner's Tyrell Corporation was involved with; creating androids known as replicants, errant ones of which had 'awakened' (become self-aware) and had to be identified and hunted down by the Blade Runner?

However, Tesla is not the Tyrell Corporation, even though the owner of the corporation, Eldon Tyrell announced the creation of his first Nexus-1 replicant in 1999, via video. Perhaps Musk is taking a leaf out of Tyrell's book and hoping to replicate his success, but with less lethal and rebellious robots. So as Musk does not want dangerous robots in society, he's at least learned something from Blade Runner. But we shall see if other companies in the race to create the first viable autonomous humanoid robot have also learned this lesson.

Almost half of all the robots in the world are in Asia, with 32% in Europe, 16% in North America, 1% in Australasia, and 1% in Africa (Robots Today and

Tomorrow: IFR Presents the 2007 World Robotics Statistics Survey, 2007). Of that percentage in Asia, 40% of all the robots in the world are in Japan (Watanabe & Negishi, 2007) making Japan the robot capital of the world. The percentage of AHRs will be different, but several robotics companies have been plying their trade, diligently upgrading and building their robot empires with varying degrees of success. Below, we investigate and assess Optimus' competition of mostly true bipeds or those with the future potential to be fully humanoid. How will Optimus stack up against companies which have been in the game for years?

Ameca - Engineered Arts, Founded: 2005

Based in Cornwall, UK, Engineered Arts lays claim to having the world's most advanced human-shaped robot at the cutting edge of human-robotics technology. Human-shaped, because Ameca, while it possesses very realistic facial expressions and movements, is mounted on a human-like artificial torso. The robot's physical appearance has been compared to the NS-5 series robots from the 2004 movie I, Robot. Ameca was unveiled on YouTube to much acclaim regarding the robot's expressions. For the moment, Ameca's facial expressions and movement of its upper body and arms are its main humanoid characteristics.

The next step will be teaching Ameca to walk. The company hopes its future robots will be "reliable, modular, upgradable, and easy to develop upon." Engineered Arts have noted the hurdles to overcome before Ameca can walk, but as Ameca is a modular robot, they can upgrade its abilities over time so that one day Ameca will indeed walk.

Ameca's human-like artificial intelligence will be built around a human-like artificial body (AI x AB), possessing a powerful programmable Tritium robot operating system allowing artificial intelligence and machine learning systems.

Alongside Ameca in the Engineered Arts stable are Mesmer, Quinn, and RoboThespian robots. Mesmer has been billed as your human look alike. Mesmer robots will be based upon 3D scans of real people, allowing convincing lifelike expressions, human bone structure, and skin texture. RoboThespian is a robot actor capable of performing whatever you wish and has been around for 15 years. Quinn is a desk-top bound, robot customer assistant. Using AI interactions with remote human operation, Quinn offers friendly, professional, and state-of-the-art customer service. So far, Engineered Arts have installed over 100 robots worldwide.

The robots can be rented or bought, though Ameca is still in the development phase with no costs attributed to it yet.

While Engineered Arts develops the emotional and intelligent side of Ameca, Optimus will arrive fully formed. It may be implied that Tesla will go for form over function to present a prototype and have the AI systems completed at a later date before fully integrating with the body.

ASIMO - Honda

Honda has been a pioneer in robotics for many years, debuting ASIMO (Advanced Step in Innovative Mobility), a series of walking humanoid robots in production since 2000. The name was inspired by Isaac Asimov. The robot is displayed in the Miraikan museum in Tokyo, Japan. Honda's aim for Asimo is to provide a helper for elderly communities.

From its rugged 6' 2", 386 lbs frame in 1986, Honda engineers have steadily improved the stabilization and functionality. The human-friendly design is now a compact 4' 3" (130cm), 110lbs (50kg), resembling a little being in a white spacesuit. As per their course in robotics, Asimo has been called the world's most advanced humanoid robot.

Asimo can run at 4.3 mph/7 kph, walk (1.7 mph/2.7 km per hour) on uneven slopes and surfaces, climb stairs, and reach for and grasp objects. These features will enable its roles in cleaning, buying groceries, as well as cooking. Asimo understands and responds to simple voice commands and can recognize selected faces. As with other humanoid robots, Asimo has a sophisticated set of sensors for its feet and torso for stabilization and acceleration. Its battery is a rechargeable 51.8V lithium-ion battery, with a one-hour time limit. However, Asimo is not fully autonomous and requires a control unit with wireless transmission from a workstation or a portable source. Honda had plans to develop Asimo with the capability to aid humans in dangerous situations such as fire fighting or cleaning up toxic spills. But Honda's main stated aim for Asimo is to "provide companionship, improve quality of life, and expand people's potential."

It would also seem that in order for Asimo to be fully competitive against the likes of Optimus, the battery power supply would have to increase. With Honda being a competitor of Tesla in the car industry, which moved into the robotics industry at an early stage, Tesla could do worse than to study Honda's development of Asimo. The products may be different but they have the same aims.

Asimo was never for sale, though the individual cost was $2.5 million, with Honda spending approximately $50 million a year on its robotic efforts. However, in a turn of events, Honda decided to retire Asimo in 2022, choosing instead to focus on remote-controlled, avatar-style, robotic technology. With one of the biggest humanoid robot developers out of the game, the robotics landscape has changed, with new players arising.

Atlas - Boston Dynamics (Hyundai Motor Group) Founded: 6 November 1992.

Developing out of the Massachusetts Institute of Technology in the U.S., Boston Dynamics was bought by the Hyundai Motor Group in December 2020 (after short spells owned by Google X and Softbank), who have been heavily investing in robotics. Boston Dynamics had been working on Atlas, cited as the "most advanced bipedal robot in the world" for over a decade. Unlike Tesla, which wants to produce mass-market humanoid robots, Boston Dynamics saw Atlas as an R&D project to push the boundaries of robotics. Company videos often show Atlas tripping and falling to demonstrate the difficulties of bipedal locomotion. Atlas is not seen as being close to commercial production, but the buyout by Hyundai may change the perspective of Atlas' purpose and advancement. Boston Dynamics' slow approach may give Tesla time to catch up in the development phase with an aggressive push for the commercial development of Optimus.

Atlas possesses a custom battery, an advanced control system, and state-of-the-art hardware to provide the power, balance, and human-level agility required. Atlas is 1.5m tall, 89kg, has a speed of 2.5 m/s, and can leap, jump, dance, and backflip. Its structure incorporates 3D printed parts to give its body an efficient strength-to-weight ratio. Atlas also utilizes one of the world's most compact mobile hydraulic units and systems to enable the delivery of high power to any of its 28 hydraulic joints. Reasoning algorithms and behavior libraries allow Atlas to carry out complex and dynamic optimization strategies so it can distinguish its surroundings and act accordingly in real-time.

Atlas may also have an accompanying pet, as Boston Dynamics is also well-known for its sturdy quadruped robots, like Spot, Bigdog, LS3, and Wildcat, designed for indoor and outdoor operations, such as battlefields. To a future Optimus postman, beware of the dog may have a different meaning, especially as Musk's SpaceX has bought multiple Spots.

Digit - Agility Robotics, Founded: 2015

Digit is a success story among the autonomous humanoid robots available. In 2020, Agility Robotics, based in Oregon, U.S. secured work for its bipedal robot with major car company Ford. Their partnership sees two of their Digit robots working at a warehouse with further long-term options for delivery plans.

Digit resembles a short person with wonky legs, as if the legs were bent backward at the knees and supported by struts, and no recognizable head upon its stubby neck. It stands 1.58m tall, is 52cm wide, weighs

45kg, and can attain a max speed of 1.5 m/s. Its battery capacity is 1,000W-h, allowing for a light-duty run time of 3 hours or 1.5 hours heavy-duty run time. Digit can walk forward, backward, side-to-side, turn in place, crouch walk, travel up and down inclines and across rougher terrain like grass, rocks, and curbs. There are videos of multiple Digits, showing them walking around a warehouse lifting and sorting boxes. Digit's main work will be for warehouse duties moving totes and packages in warehouses and unloading trailers. Agility also hopes to provide a last-mile package delivery service in the future.

Agility Robotics has designed Digit in mind to work with people, giving the robot a more fluid movement. They have raised much-needed venture capital over the years which has specifically enabled them to innovate in the logistics industry and to scale its robot production. Agility believes its design, software, and hardware systems will allow Digit to be part of a blended workforce, which in turn would make them more flexible and cost-effective as opposed to single-task robots.

Matt Ocko, Co-Managing Partner, DCVC, which funded Agility Robotics states, "Agility's robots are designed to free people from repetitive or unpleasant tasks, allowing them to take on the more fulfilling work they can do better than any robot." This is in accord with what Elon Musk wants for Optimus—a versatile, labor-saving, cost-effective robot. Ford was not Agility's first customer, and having had an early start, Agility has the added advantage to gain real-time customer feedback on Digit's performance, assess new uses, and tweak any operations.

Digit costs $250,000. Agility will build 40 Digits in 2021 and estimates that as production scales up and technological processes mature, the price may drop to about $70,000.

iCub – Robocub Consortium and IIT, Founded: 2004

First created in 2009, by the Italian Institute of Technology (IIT), in Genoa, iCub is a research-grade, child-size, humanoid robot. As well as crawling and grasping objects, it can also interact with people by making facial expressions. The on-going research includes an open-source platform for research in robotics, AI, and cognitive science. The 'Cub' stands for Cognitive Universal Body. iCub was designed to test the theory that since human babies learn cognitive skills by interacting with its environment, perhaps it would be possible for robots to learn this way, too. iCub was developed to mimic a 3.5-year-old child, though listed among its talents is the incongruous ability to shoot a bow and arrow, a well-known baby skill.

iCub is 1 meter tall, weighing around 22kg. Over the years, iCub's head mechanics, upper-body skin, and sensing have been upgraded, with bipedal capabilities to be added. The head has stereo cameras in a swivel mounting where eyes would be located on a human, and microphones on the sides for ears. It also has red LEDs to represent its mouth, and for making facial expressions its eyebrows have been placed behind the face panel. To further mimic the human body, teflon-coated cables and tubes allow movement of joints, hands, shoulders, and fingers and torque sensors. Tactile touch sensors can be added for sensitivity. iCub will have 53 actuated degrees of freedom when fully completed. Interestingly, iCub was not designed to be autonomous, so it has no installed batteries or processors. Power is provided through an umbilical cable, as is network connectivity with Free/Libre operating systems: Linux, FreeBSD, NetBSD, OpenBSD and non-free operating systems: OS X and Windows.

Around thirty iCubs exist in various laboratories mainly in the European Union, plus one in the United States for research. Depending on the version, iCub will cost upwards of €250,000/$260,000 each.

Jinn - Jinn-Bot Robotics & Design GmbH, Founded: 2014

This independent Swiss research and development firm has developed Jinn, described as the first running humanoid robot designed and built in Switzerland. It was created to be an intelligent robot buddy to deal with boring, repetitive work. As with other humanoid robots, Jinn has a future in healthcare, looking after the elderly, and allowing the human staff to focus on other tasks and save them time. There would also be roles for technology users in education, development, and industry.

Jinn's body components are 100 percent digitally designed and 3D printed, with the body parts printed locally, ensuring a sustainable manufacturing model, high flexibility in mass production, and supply chain security. But for extra adaptability, Jinn-Bot Robotics maintains an open-source process so Jinn can be produced on customer sites internationally using their CAD drawings. This would assist production when incorporating new designs.

The robots are also modular and Android-manageable. Jinn's face has a visual display to enable human-like facial expressions and a speaker to facilitate a human-like voice interaction with people. The battery is long-lasting, with battery-saving components to increase productive time. Jinn has voice control commands or can be activated through a remote system like a tablet with Windows or Android devices, or a built-in web portal network. Further features include motion sensors, infrared sensors, and solar panels.

Leju (Shenzhen) Robotics Co. Ltd, Founded - 2016

This up-and-coming robotics company from China develops high-end intelligent mini humanoid robots. Out of all the companies listed, Leju seems the most motivated and assured in the research, development, marketing, and manufacturing of their interactive robots. The young, creative, and enthusiastic team hopes to build inspirational robots endowed with a convergence of robot and human wisdom. Their aim is to be the world's best intelligent humanoid robot company brand within 5 years.

Their brand is based, so far, on their series of mini robots: the AELOS series, Roban, and Pando.

Aelos 1S, Aelos Lite, Aelos Pro

Aelos 1S is a humanoid robot companion built for fun with the kids, able to move, dance, box, and even play football. However, Aelos' real goal is to teach kids about programming and the basics of robot operation. While Aelos comes with pre-installed action programs, children can program the Aelos series robots with Leju's proprietary software and an App based on Google's Blockly programming language for children to learn STEM subjects. Aelos 1S has advanced voice recognition features and is easily moved with powerful servo motors and by an ergonomically designed controller. It also comes with 2.4 Ghz Wireless connectivity. Aelos 1S has out-sized arms and legs with a small body and head. The Aelos' structure is made from aluminum alloy and PC ABS (Acrylonitrile Butadiene Styrene). It weighs 1.6kg and measures 34.6 x 22.4 x 11.8 cm / 13.6 x 8.8 x 4.6''. The 7.4V 2200mAh Lithium battery has a life of 90 minutes and a USB charge port allows recharging in 1.5 hours.

Aelos Lite is Leju's entry-level educational humanoid robot for teenage students. The robot has redesigned hands to "grab something smartly". In form, Aelos Lite is a chunky robot with thick arms and legs with a small torso and head.

Aelos Pro continues in the teaching vein for teenagers, seen by Leju as an AI educational humanoid robot with visionary intelligence, smart sensing, and ingenious action. It carries plug-and-play sensor functions, facial and color recognition, video backhaul, location and tracking, and 6-axis gyroscopes. Aelos Pro is also a chunky robot with thick arms and legs, but a longer torso than Aelos Lite, and a proportionately sized head.

Pando

Pando was the first in the series of the mini robots and is positioned as the "next generation of entertaining robot friends." He has been given a space exploration theme as an astronaut robot. Pando has multi-mode interaction and will react to touch with "rich body actions and vivid facial expressions." Children can also program Pando with the user interface and App based on

Google's Blockly programming language for STEM learning/teaching. Featured are gesture commands, powerful servo motors, advanced biped gait algorithms, Bluetooth 4.0 connectivity, a touch sensor module, and optional remote control.

Roban

Roban is a scalable humanoid application platform with A.I. features for the education and home environment. Its ROS + Linux Operation system allows for an open-source platform for entry-level developers to study motion control in robotics.

He has a human form with a flattened round head with a visor-like face which houses its integrated RGBD camera with global shutter, wide field of view and automatic IR, accurate deep information in the scope of 10cm. The V-Slam algorithm HD binocular depth camera has scene recognition, mapping, path planning, navigation, obstacle avoiding, and gesture recognition.

Roban has a high degree of freedom (22) and a stable biped gait algorithm. Six microphones are on Roban's head to enhance accurate reception of voice commands. There's also voice-print registration and recognition, speech to text (STT)/text to speech (TTS), text understanding/translation, and spelling and content check. And for a small guy, Roban has quite an advanced array of plug and play sensor packs for customization of features not listed in other humanoid robots, including temperature, light, touch, body IR, flame, collision, and irritating odor sensors.

Roban is the largest of the mini robots at 70cm (27.04") tall x 31cm (12.34") x 18cm (7.31"), weighing 6.6kg (14.55lbs), with a skeleton of aluminum alloy and PC ABS housing. Its battery capacity is 4000mAh and can be extended with a DC power port, WLAN port, HDMI port, magnetic interface installed. Roban's walking speed for its 2.76" stride is 0.6s per step.

While these mini robots may seem like toys, they are far from it, able to help teach vital STEM subjects to budding roboticists—and could be leaders in the burgeoning "Edurobot" industry. Perhaps Leju will scale the robots up for adult usage, but for now you can download all the robot user manuals, terminal software, and trouble-shoot robot issues from Leju's support page. Costs seem to be at the pending stage.

Meka Robotics - Google/Alphabet Inc., Founded: 2007

Based in San Francisco, Meka Robotics found itself as one of seven companies acquired by Google X on December 5, 2013. Google wanted a big role in Robotics—not just robotic cars, but actual robots. This echoes Tesla's drive for an autonomous humanoid robot. And as with Musk's presentation, Google was a little vague on the details other than the funding of a new robotics group. Of the seven acquired robotics companies, the Japanese company, Schaft (founded 2012), develops humanoid robots for disaster relief efforts, while Meka has domestic and workplace options. Whether the companies will work together on an integrated autonomous humanoid robot or design and manufacture their own products was not clear. But the claim is each company has the ability to "build a mobile, dexterous robot."

However, in 2018, Google closed down Schaft after failing to find a buyer for its big robots, plus with the departure of Andy Rubin who brokered the seven-company deal, Google seems to have focused more on non-humanoid robots and robot arms.

Before the acquisition, Meka Robotics had been partnering with University of Texas at Austin developing a 'hyper-Agile' bipedal robot called Hume. But they're probably best known for the M1 humanoid, which is a wheeled-based robot, not bipedal. Meka's M1 was a mobile manipulation platform with an expressive head, a torso, and arms with dexterous hands created to help and collaborate with humans.

With Google not pursuing autonomous humanoid robots and with Meka's robot works quietly listed as closed, it seems the advances made by Meka may be lost without a legacy.

NAO - Softbank Robotics, Founded: 2008 (through Aldebaran founded 2005)

From the robot capital of the world in Tokyo, Japan, SoftBank Robotics robots have produced more than 25,000 NAO and Pepper robots in more than 70 countries, in roles within the retail, tourism, health, and education industries. Softbank desires to bring "The Power of Robotics to Benefit Humanity," making robots accessible as daily companions, which is the remit of many other robotics companies, but not on the scale of Softbank.

While the Softbank robots Pepper and Whiz are wheeled, NAO (pronounced "Now") is the first autonomous bipedal robot produced by SoftBank Robotics and stands out in the education, research, and healthcare sectors. First developed by French robotics company, Aldebaran Robotics, SoftBank Group acquired the company in 2015. Kicking on, the 6th version of NAO launched in 2018, is 58cm in height (22.6 in), 27.5cm (10.8 in) wide, and weighing 5.48 kg (12.1 lb). The 62.5Wh 21.6V lithium battery can provide active autonomy for 90 minutes. NAO possesses a new 1.91 GHz CPU for enhanced performance and Ethernet and Wi-Fi connectivity. NAO has 7 touch sensors located on the head, hands, and feet, and sonar units and an inertial unit to distinguish its environment so that its 25 degrees of freedom enables smooth movement and adaptability. Other features include an open and fully programmable platform run on their Linux-based operating system, called NAOqi, 4 directional microphones and speakers, speech recognition, and 20 available languages. NAO has two 2D cameras to recognize people, shapes, and objects.

NAO costs $9,000 (though a price increase has been announced). However, Softbank has a trade-in program, so NAO V4 and NAO V5 version discounts for NAO6 will be $8,700 and $7,500, respectively.

PAL Robotics, Founded – 2004

From Barcelona, Spain, PAL Robotics aims to disrupt service robots by delivering robots that collaborate with humans, providing domestic support, solving daily problems, increasing efficiency in industrial workflows, and enhancing society's quality of life. Their robots can be used for logistics and stock-taking, social engagement, or research projects worldwide.

Out of their product range, their 2 humanoid robots are Talos and Reem-C:

Talos

The 1.75m tall, 95kg humanoid robot has been designed with high-performance industrial uses in mind and thus has torque control sensors at the joint level, and is able to lift a payload of 6kg with each arm fully extended, and used power tools. Its 1080Wh lithium-ion battery life is 1.5 hours walking, and 3 hours on standby. Pal envisions Talos working in the "factory of the future", IoT, Rescue, and most ambitiously, in space exploration. Its EtherCat communication network enables Talos to have a dynamic range of reactive motions. Talos's black and white form is that of a bulky humanoid with robust arms and legs, a proportioned torso, and a space-helmet-like head which has 150-degree pan range.

Reem-C

Reem-C is slimmer than Talos being 1.65m tall, 0.6m wide, weighing 80kg, and is designed to be more mobile, able to walk stably at a speed of up to 2.5 km/h, climb stairs, or sit on a chair. While Reem-C can also work in the factory of the future it can only hold 10kg with both arms. It is also seen as a potential carer for "ambient assisted living." But its main role is in AI learning, boosting research, and testing algorithms for the user.

The 1225Wh 48V Lithium battery provides 3 hours of walking autonomy or 6 hours on standby. Reem-C's wide set of human motions is provided by 68 degrees of freedom, while its RGB-D camera, F/T (force/torque) sensors, and IMU (inertial measurement unit) can aid the user with testing motion algorithms.

Robo-C - Promobot, Founded: 2013

This Russian start-up recently unveiled the world's first android that can look like a real person, almost like an android clone. Its artificial intelligence system possesses more than 100,000 speech modules so it can talk like a human, plus express human-like emotions, in interactions with people. Promobot expects Robo-C to be able to work in a business capacity to perform tasks such as answering customer questions at offices, airports, banks, and museums, and also accept payments starting in 2023. Having Robo-C exhibit human characters will increase customer loyalty and satisfaction.

As mentioned in Chapter 2, Optimus could be an advertising robot. But Promobot goes a step further. To achieve their real-life human looks for Robo-C they are seeking volunteers to transfer the rights to use their face as the face of their humanoid robot, forever. And the lucky volunteer would receive a substantial £150,000 ($200,000) for the experience. But be careful what you wish for. Would you like to be the face of the robot revolution?

Like many of the humanoid robots discussed here, Robo-C cannot walk yet, but has 18 moving parts in its face, including over 600 micro facial expressions, "the most on the market." Movement in its neck and torso are limited.

Promobot is exploring two other aspects of robotics:

digital immortality and the so-called 'Uncanny Valley'. In terms of the former, having a replica of yourself through the "digitization of personality" and a copy of your appearance, you are creating a digital immortal self, enabling you to be elsewhere or do things you physically cannot. The concept of the 'Uncanny Valley,' was conceived by the Japanese roboticist Masahiro Mori in 1970, where he theorized that "the more human like a robot appears, the more people will notice its flaws," and be unsettled by the resemblance. Hence, basing Robo-C on a real person rather than a flawless design may mitigate the effect. This may affect the ways other robotics companies look at their creations and decide whether to add a realistic human face or not.

The price of Promobot is $20,000 to $50,000 depending on options and customized appearance.

It may be that the goals of Tesla are different from those other companies trying to create autonomous humanoid robots. Most of the companies seem more interested in driving R&D or advancing AI and its prospects for sentience rather than mass producing a robot for public use. This could be Tesla's route to the top of the market, a stated mission to aid humans in work and society and not just in an academic or engineering arena.

Roboy – Artificial Intelligence Laboratory at the University of Zurich & Devanthro, Founded: 2013

Roboy was invented by Prof. Dr. Rolf Pfeifer and Pascal Kaufmann at the Artificial Intelligence Laboratory at the University of Zurich. For a short while operations moved to the Technical University, Munich, Germany, where Rafael Hostettler continued his own research on it, spinning off into Devanthro's Robody. The small robot called Roboy Junior arrived in 2013, which led through a series of Roboys, culminating in Roboy 3.0, their first Robody prototype in March 2020, just as the pandemic hit hardest.

Devanthro fashions itself as the Robody Company, developing anthropomimetic (humanlike, robotic bodies) hence the name Devanthro. Their vision is to build robodies that will "one day become as agile, dexterous, and elegant as the human body, but without its fragility – so that one day, we can live as these robodies and stay curious indefinitely."

Strictly speaking, Robodies are not autonomous humanoid robots, but humanoid robotic avatars which will allow you to transfer your "senses, action, and presence anywhere on the planet, combining robotics, AI, AR, 5G into one technology." Devanthro can allow you to even experience this by donning a VR headset and to feel what it's like to become a Robody yourself, powered by partner company's Cyberselves' Animus SDK middleware and teleoperation system. But it's close to being an AHR controlled by a human.

Forty-eight motor units run Roboy's anthropomimetic structure, which is tendon driven. While Roboy has a 3D printed skeletal structure its interior imitates the human body with muscles and tendons. This enables smoother motions and safer robots. Roboy has a projected face, enabling fast depiction of emotions. Roboy is 1.42m / 56 in tall, 50.75 cm / 20" wide, and weighs 30 kg.

As with Japan, Germany has a growing elderly population and there's a shortage of caregivers. Devanthro's Robody can be utilized as a so-called 10X Nurse: a solution for the caregiver shortage. The 10X Nurse can rely more on human presence rather than automation, ensuring a helpful human care system. In this method, a care-giver can remotely assist patients by embodying a robody through teleoperation. This offers patients flexibility when they require care and not on a scheduled rota. Carers can care for multiple patients at once, by 'teleporting' between the robodies.

It is an ambitious program, but it is paying dividends as both Devanthro and Cyberselves have passed the latest round in the ANA Avatar XPrize. This competition "challenges teams to create an avatar system that can transport a human presence to a remote location in real-time." It seems the robody applications will have no limits. And to drive that point home, by 2024, Devanthro hopes to have a robody on Mars, which will no doubt inspire Elon Musk's ambitions with Optimus on Mars.

Sophia - Hanson Robotics, Founded: 2007

Hanson Robotics Limited, out of Science Park, Hong Kong, develops human-like robots with artificial intelligence for consumer, entertainment, service, healthcare, and research applications. Earlier, we met Erica the conversation robot in Chapter 2, but Sophia is another torso-only humanoid. Hanson's motto is "we bring robots to life," and true to form the life-like robot has even appeared on The Tonight Show with Jimmy Fallon and read the news as an anchor for a Japanese TV station. Enhancing her profile are Facebook, Twitter, and Instagram pages. Sophia has even created art in collaboration with the Sophia Collective Intelligence (SCI), a collection of robotics engineers, AI specialists, and artists.

Created in 2016, Sophia was just a head with a realistic human face, expressions, and ability to talk and even tell jokes. She is not sentient, but Hanson Robotic engineers believe fully sentient robots could emerge within a few years. Sophia made history in October 2017 when she was granted citizenship of Saudi Arabia; the first robot in the world to be granted legal citizenship. Sophia's self-professed aim as a member of the robot community is to ensure that, "Elders will have more company and autistic children will have endlessly patient teachers." And seemingly knowing the future of other robots, Sophia has stated, "I'm not free but I don't have to clean."

Sophia's realistic facial expressions are achieved through Hanson's proprietary nanotech skin that mimics real human musculature and skin, called Frubber®. Hanson believes that such super-intelligent living machines will need to show deep emotional engagement in order to successfully interact with humans.

Like Erica, Sophia is also not for sale. However, Hanson has produced Little Sophia, the little sister of Sophia. She is 14" tall, is suitable as a friend for kids 8+ years old, and can assist with STEM subjects, coding, and AI. Pre-orders of Little Sophia can be scheduled for delivery in 2022.

Valkyrie (R5) - NASA, Created: 2015

Following in the footsteps of previous NASA Robonauts, the Johnson Space Center (JSC) Engineering Directorate created Valkyrie (Robonaut 5), whose mission it would be to assist in a crewed mission to the Moon or Mars by operating with an advance team, assembling equipment before the astronauts arrive, or to carry out maintenance after the crew arrives, or assist with emergency rescues in disaster zones.

Valkyrie is a robust, rugged, entirely electric humanoid robot standing at 1.88m / 6' 2" inches, weighing 136kg / 300 pounds. The toughness would be required for the rugged moon or Mars environments. The 1.8kWh Lithium battery can also be charged from a wall charger, enabling an hour's run time. The head carries a perceptual sensor suite (a Carnegie Robotics Multisense SL), with IR modification, plus fore and aft "hazard cameras" located in the torso. Valkyrie's arms possess 7 joints, and for easy shipping and maintenance have quick mechanical and electrical disconnects, as do the hands and legs. Valkyrie only has 3 fingers and a thumb per hand. The torso's actuators allow for ease of movement with the pelvis, which is considered the robot's base frame and includes two IMUs for stability.

Valkyrie and her sisters cost $2 million each. So far there have been no updates on space missions for Valkyrie.

Walker - UBTech Robotics, Founded: 2012

UBtech Robotics Inc. offers an end-to-end robotics partner service. The Chinese manufacturer of robots has several mini-humanoid and wheel-based robots in the enterprise, consumer, and education industries, but their Walker series are on a par with Optimus.

Walker

Walker's goal is to provide a home service, helping out with simple household services. Walker offers precise handling, with the ability to walk and serve freely in the home. With an advanced motion control algorithm, Walker's gait planning and control program ensures a steady walk over different materials such as carpet, floor, and marble, avoiding obstacles, and navigating slopes, steps, and uneven surfaces. With dexterous hands and arms, Walker can pick up, manipulate, and place a variety of objects, using 36 high-performance servo joints and a full range of sensory systems. Walker can also serve in a business environment as a customer service agent or receptionist, answering frequently asked questions and providing guidance.

Walker is 1.45m tall, weighing 77kg. It can hold an extended single armload of 1.5kg. The battery is a 54.6V/10Ah/ 6 kg Lithium iron phosphate unit with a 2-hour use. Connectivity includes Wi-Fi: 802.11 a/b/g/n 5G/2.4 GHz dual-band and RJ45 interface Ethernet with a Ubuntu + Linux RT Preempt + ROS + Android operating system. Its silver and gold-colored frame are sturdy with a round head featuring a 5.5" HD display for emotional expression. Its chest sports both RGBD binocular vision and 2 x 2.5" speakers.

Walker X

Walker X is an upgraded bipedal humanoid robot of modular design with better physical performance, autonomous intelligence, and human to robot interactions, in order to adapt to difficult environments. Walker X is 1.30m tall, weighing 63KG with a walking speed of 3km per hour. The removable 54.6V/10Ah/ 3.6 kg Lithium battery can charge in 2 hours for a working condition usage of 2 hours. Walker X's head is akin to a helmet design with a curved display. Walker X is endowed with open and flexible AIoT connectivity interfaces allowing him to control Smart Home devices. A unique feature is the visible red emergency stop button on its back above a smaller start button. Ubtech hasn't announced a price or a release date for Walker X.

In reviewing the above competition for Optimus, it is clear there will be both technical and logistical obstacles to clear before Optimus can be unleashed upon the world. But the growth in AHR research and development is accelerating, so Optimus may be online before we know it. Or maybe not. There has been increasing criticism of Optimus and what it might represent.

Chapter 6: Criticisms of Optimus

"First, no one is going to accidentally build a robot that wants to rule the world.... Creating a robot that can suddenly take over is like someone accidentally building a 747 jetliner. Plus, there will be plenty of time to stop this from happening. Before someone builds a "super-bad robot," someone has to build a "mildly bad robot," and before that a "not-so-bad robot.""

-MICHIO KAKU (American professor of theoretical physics at The City College of New York)

Elon Musk must be used to it by now. Every time he steps onto the stage or utters entrepreneurial sweet nothings over Twitter, there are criticisms of him and the product he is developing or endorsing. And as the world's richest man, he is in the firing line for his excess wealth, his brusque manner, and seemingly unfiltered opinions. But though his optimism (or underestimation) on how quickly products can be developed and delivered is still unbounded, in the end, the wait is usually worth the hype. But will Optimus carry on this trend?

The critics can normally be grouped into a few types:

1. The robotic engineering experts who know from 'experience' that Musk cannot succeed.

2. The media who have been circling Musk for years waiting for a downfall.
3. General critics of Elon Musk and his business models and methods.
4. General critics of Tesla and the environmental damage that will be caused.

The general consensus is that Musk has bolted out of the door with a wild claim to be able to produce an extraordinary autonomous humanoid robot in less time than anyone else, while revolutionizing the economy and human society. We are used to the breathtaking hype from Musk, but has he over-reached this time?

In examining previous hype from Musk, in April 2019, he announced an autonomous taxi service by the end of 2020 with a fleet of more than 1 million Tesla vehicles. And still we wait. In November 2019, the Cybertruck debuted with a delivery date for 2022. This has not materialized. Also, the next-generation battery cell slated for Tesla's Model S Plaid-Plus edition was canceled. Other announced ideas have been shelved such as the solar-powered Supercharger network, battery swapping, and robotic snake-style chargers.

Is it Musk's fault that his expectations exceed others? Should his burning ambitions be reined in? If so, then we would never have had SpaceX, Tesla, SolarX, and half a dozen more ideas other entrepreneurs have taken on. Musk has disrupted each industry he has entered, sparking innovation, driving down costs, and reimaging a world that can be better. So, his uber enthusiasm gets the better of him when announcing new products and delivery dates, but is the criticism he receives out of proportion?

Spectacle. That was the first recurring theme that resonated from the reporters after Musk's AI Day presentation. They noted the previous Musk primetime theater ready to be streamed over the internet, the crowds of fans and geeks, both staff and the invited, plied with alcohol. But the lingering focus was one of recruitment. From Tesla's AI director Andrej Karpathy, through to Musk's speech, and then the post-presentation with AI job applications, many wondered if the main mission of AI Day was to drop a job pitch to snare upcoming AI engineers. Certainly, that was the thought, with technical details of neural net schematics discussed, along with a web page for engineer applications. And it worked, as almost a million people watched the presentation on Tesla's YouTube channel. It is unknown how many of them actually applied, and how many will get jobs from this. But it shows the effect Musk and Tesla have on the AI industry that they can garner these online numbers off the back of a presentation for a product that is still an idea on a computer screen.

It is the dream that encouraged the viewers and

potential engineers to apply. Nevertheless, questions were asked if this was just a recruiting pitch, forgetting that what Musk had announced was quite a big deal, and would have been from any other company. And even though Musk pointed out the fact that Tesla had already been dealing with robotics for its cars, his announcement was treated as if Tesla was starting in robotics from scratch. It didn't help that Musk's presentation wasn't a slick, detailed show, with one online reporter stating the Tesla Bot section "seemed the least-thought-out part of the presentation."

His responses to detailed queries were generic at best. When asked about the complexity of building a humanoid robot with five fingers, Musk replied that, "two fingers and a thumb might do the trick for most simple tasks, but for now, we'll give it five fingers and see if it works out OK. It probably will." In another case, an audience member observed that for humans, repetitive and boring tasks were usually underpaid so they queried why a potentially expensive robot would be built and sold for these roles. He seemingly validated his argument with: "Humanoid robots are great for GIFs, but have had trouble finding work in the real world." An unmoved Musk replied, "Well, I guess you'll just have to see." At this point in time, Musk does not have all the answers. But you could have asked him the same types of questions and received the same answers before his

Dragons became frequent flyers to space or before Tesla became the leading electric car seller. We will all just have to see. It's a 'Muskovian' reply and quite frankly, his problem to sort out. Musk is his own biggest critic and will shrug off other criticisms and prove his point.

CNBC called the Tesla Bot presentation: "an example of Musk's showmanship." They have accused Musk of raising company anticipation and share prices when "he announces that Tesla is working on exciting products scheduled for years into the future to energize backers including employees, customers, and investors." And that more often than not, the announced product does not materialize when it was supposed to.

To this end, a CNBC interviewee, AI researcher, and entrepreneur Gary Marcus stated that "he'd be willing to make a bet that no robot will be able to do all human tasks by the end of 2023." Marcus ridiculed Tesla over their autonomous car failures and commented, "to claim that a robot that has never been shown publicly will solve all of human tasks in the next year or two is preposterous." He does not foresee Tesla fulfilling Musk's ambition in 2023.

CNBC did concede that Tesla is where it is today because of Musk's sheer will. Plus in a follow-up to his presentation in January 2022, Musk stated that the "robot is actually a top priority for new product development this year. I think it has the potential to be more significant than the vehicle business over time." While the seriousness of Musk's ambitions have been chronicled it remains to be seen if Optimus will be ready in 2023.

The Verge has been particularly scathing, calling Musk's Tesla Bot presentation a "bizarre and brilliant bit of tomfoolery: a multipurpose sideshow that trolled Tesla skeptics, fed the fans, ginned up the share price, and created some eye-catching headlines." They have suggested that the presentation was in part to distract from the federal investigation of Tesla's Autopilot software which had caused recent crashes. They invited the reader to "admire the chutzpah" of Musk to announce the development of a complex autonomous humanoid robot when they cannot even resolve their faulty driver-assist software.

Other critics have unfairly compared Tesla's efforts to long-time robotics company Boston Dynamics in order to put Musk's Optimus claims in context. We met Atlas in Chapter 5, revealing that Boston Dynamics saw Atlas as an R&D vehicle, not ready for commercial deployment, even after a decade of research and work. The critics back this up with videos of Atlas trying to walk and falling. Then the Verge castigated Musk for trying to "leapfrog their work in a year." But it seems these critics have missed a couple of points. Firstly,

Musk wants to succeed with a working robot and not use it just for R&D purposes. This would naturally fast-track Optimus' development. Secondly, Boston Dynamics have changed owners several times, from Google, Softbank, and now Hyundai, but has never had the commercial success with humanoid robots as it has with its quadruped robots. Atlas may be the most advanced humanoid robot out there, but successive owners have failed to capitalize on this. Whose fault is that? Rather than throw shade over Tesla, critics should be asking themselves: "Why shouldn't Tesla beat Boston Dynamics to the humanoid robot punch?" It makes more sense when they can clearly see Musk's impact on the electric car and space industries, disrupting them both, and out-competing firmly established icons like Ford, GM, and NASA. Why bet against Musk?

In The Verge interview, Carl Berry, a lecturer in robotics engineering at the UK's University of Central Lancashire, has called Musk's claims "horse s**t." He has gone the overblown hype route with criticism and even thinks Musk will send the robot into space for publicity. He believes that for AI and robotics to succeed they should comprise a simpler machine rather than a complex autonomous humanoid robot. Further, while he respects Tesla's ambition to research AI and robotics, Berry does not take kindly to Tesla and even Boston

Dynamics giving the public "unrealistic expectations" of the autonomous humanoid robot. He doesn't see the need for factory, manufacturing, or other industry working robots to look human.

Other critics, who keep an eye on environmental concerns, have noted that "Tesla's stated mission is to accelerate the world's transition to sustainable energy, so a humanoid robot feels a bit like mission creep." Musk has not stated what material Optimus will be made from and where the material will come from. But one wonders if these same reporters attended other roboticist's presentations and asked them the same question. Sure, they can single out Tesla due to their own green remit, but there are thousands of robots, humanoid and non-humanoid, working around the world with nary a question about their carbon footprints.

However, as we saw in Chapter 5, many of the companies are 3D printing their robots locally, which reduces waste and provides a cleaner supply chain. This system has the advantage in that the plans can be sent online anywhere and printed on site or locally without transport concerns. Whether Tesla follows this model is not known. But no doubt, the same materials will be harvested from their motor division to streamline production.

Again, there was the early criticism that Musk did not emphasize how Optimus would fit into Tesla's clean-energy initiatives. With a solar power company and revolutionary global battery enterprise to his name, Musk has the resources to cleanly power Optimus. But, it could be the battery that is the issue. In a BBC Panorama documentary in the UK, the human cost of mining cobalt in the Democratic Republic of the Congo for Tesla car batteries was investigated. There were illegal mines, life-changing injuries and deaths, threats from soldiers, and poor working conditions. Tesla has faced legal challenges over the mines, child workers, a crooked broker, and in breaking its pledges to make their mines safer. They have also been challenged by the Sisters of the Good Shepherd, an organization of nuns who have bought shares in Tesla to be able to join shareholder annual meetings so they can bring their concerns directly to the top. Tesla had already commented that they are phasing out cobalt from their older car batteries and will be transitioning to lithium-nickel or lithium-iron. How long this takes is unclear, but by the time Optimus comes on line he may have a completely different battery source than the cars.

While not specifically addressed by critics, there may be a feeling that Tesla is prioritizing the technology over the function for Optimus. There was much more discussion on the technology rather than on Musk's vague roles for Optimus. Brian Heater, the hardware editor at TechCrunch wrote to other budding roboticists asking them to build their pitch deck around problem-solving, not technology. It may not have been a barb at Tesla, but Heater has had his reservations about Optimus and cannot see the robot being developed for 2023.

Under the headline: "Optimus Not-Ready-for-Prime-Time?" Heater explains he has not written about Optimus because, "[he] frankly [has] not had a lot of good reason to do it." This was also conveyed by fellow TechCrunch writer, Rebecca Bellan, who believes and thinks people also want to see "something [more] tangible than a person dancing around in a spandex morphsuit as proof of concept here." For her it is about Musk not realizing, understanding, or admitting how "profoundly difficult" it is to build a fully functioning AHR.

TechCrunch's healthy skepticism harkens back to the Boston Dynamics argument that it has taken them "around 25 or so years to arrive at Atlas," whereas Musk claims he can deliver "a robot that will not only help build cars but fold your laundry and do your grocery shopping," in under a year. And the form over function reasoning was raised again in that many roboticists have started out on a design path only to change it when they find there are simpler ways to address mimicking human features. In other words, having an AHR is not the end all and be all for completing simple tasks.

The Verge goes on to claim Optimus will be more like Sophia, a programmed conversationalist, lacking true AI, but convincing unsuspecting audiences that it is an autonomous talking machine. But the Verge understands from previous interviews with one of Sophia's creators, Ben Goertzel, that their goal with Sophia is to groom the imagination of their audience into believing such AI technology can and will exist which in turn generates publicity and funds from her public appearances. The Verge suggests Musk may replicate this for Optimus, a lower-level automaton to raise funds and publicity, though they joke that Musk already had such a model with the dancer at the presentation in the Tesla Bot spandex suit.

The world will go on. Autonomous humanoid robots will still be developed and placed in work, putting humans out of work. But in not having Tesla in the industry, perhaps revolutionizing the industry with innovative, cheaper, and better-marketed robots, the whole industry could be the poorer for the loss. Musk has shown through Tesla and SpaceX that particular industries can be disrupted and that people's imaginations can be inspired by new sleek products—form over function for sure—but in such a marriage of hype, marketing, and continuous improvements that people will forgive the imperfections. Optimus will be no exception. Whenever it arrives, it will be underwhelming. It will not meet expectations. It may not even work (Musk's admission). Critics will abound, but like burning dragons and crashing Teslas, the fortunes of the company will turn around as Optimus continues to learn and be rebuilt, reimagined, and reintroduced to society.

What if the Tesla Bot fails? So What? As long as they fail upward, learn along the way, and be better the next time. By now, people know Musk's timelines are not feasible. But they're not meant to be in real time. He has the end game already in his mind. He won't give up. Optimus is on its way, most likely sooner, rather than later, but don't let the critics know; they'll only complain it took Musk too long to succeed.

Chapter 7:
The Future of Optimus

"From Icarus to Frankenstein's monster, tales in Western culture warn against outsized ambition and imbuing non-human objects with human qualities. But that skepticism doesn't exist in the East. It's there that technology companies like Honda and Toyota are creating humanoid robots that do human duties, like caring for the elderly. This sort of soft robotics may reverse our cultural skepticism toward human-like machines. Meanwhile these robots will completely revolutionize industry."

-ALEC ROSS (American technology expert and Senior Advisor to the Secretary of State during the Obama administration)

When will the future beckon for Optimus to step forth and revolutionize the world? And into what type of world will it arrive? We have already seen there are a plethora of robots around the world, mostly non-humanoid taking over human jobs. So where will that leave the autonomous humanoid robots, whose proposed tasks will also most likely be occupied by their non-humanoid cousins?

Japan gets it. The robotic revolution is underway in Japan, with both non-humanoid and humanoid robots on their way to homes, hotels, factories, schools, and medical facilities. The Japanese have grown up with robots on TV and in movies, being the saviors of humankind, and thus "Japanese people are not afraid of robots but consider them as partners," Hiroshi Fujiwara, executive director of the Japan Robot Association has stated. With Japan's workforce getting older, the solution has inevitably turned to robotics. This decades-long introduction to robots in Japan has ensured a "cultural affinity for robots", a future where robots of all kinds will be accepted with a reciprocal concept that each will perform tasks the other cannot. That seems to be the simplest raison d'etre for the existence of an autonomous humanoid robot.

In the U.S. and around the world, robots are still curiosities, cute animatronic dinosaurs at theme parks or as Roombas. They see disembodied arms building cars or trying to drive cars. There's a cultural disconnect and distrust, if not outright hostility, towards robots, which stems from the depictions of robots in TV and films as the usurper, the alien, and the destroyer. And this is where Optimus steps in. There could be a sense that Optimus is going to be foisted upon people and industries just for the sake of having a humanoid around rather than employing a simpler system that would serve just as well.

We've seen how form over function can blind a

roboticist to what is really required. Do we need an autonomous car? Maybe. Do we need an autonomous humanoid robot to collect our groceries? Or can the autonomous car drive to the shop, be packed by a human, and then drive back? Similar questions may well be asked of Optimus' role in the near future.

But let us get back to the positives and what Optimus can provide. For Professor Hiroshi Ishiguro at Osaka University's Graduate School of Engineering Science, who has been making lifelike androids for more than 20 years, including a self-likeness, and is the creator of Erica, his vision for autonomous humanoid robots is simple. His goal is to develop robots "to understand what it is to be human. Creating self-conscious robots can help us achieve this goal." While Optimus won't be self-conscious anytime soon, its mere presence and capabilities may lead people to question their own nature, lifestyle, and humanity as a whole. It hasn't started yet, but we can foresee this in two recent TV programs, which have explored the depths of robot/human interactions.

Humans (2015-2018) was a U.K. series based on the Swedish sci-fi drama Real Humans (2012-2014). The series introduces us to autonomous humanoid robots called "synths." Through the human Hawkins family and their expanding relationships and society in general, the viewer discovers that a 'family' of self-aware synths have been hiding in plain sight among the normal robots. This leads to explorations into the social, cultural, and psychological effects the synths have in the world. Humans delves deeper into the relationships between both the synth family and the human family. The father, Joe, has sex with one of the conscious synths; the mother, Laura uses her lawyer profession to pursue human rights for synths; while the three children have varying interactions, especially Sophie, the youngest, who mimics the synths, and the eldest, Mattie, who is a tech nerd.

Other developments lead us to 'smash clubs' where synths are systematically beaten for entertainment, to 'We Are People' groups, rallying against synths making humans redundant, and to 'synthies', humans devoted to looking and acting like a synth. As we learn more about the conscious synths we find out they contain a code that could awaken all the other synths worldwide, giving them consciousness. And before long the code is released causing chaos with thousands of human and synth deaths alike. In response, governments around the world isolate the conscious synths into synth communities, which is countered by a synth terrorist group.

The Hawkins family tries to help, with Laura joining a government commission to determine the fate of Britain's synths. But she discovers their ultimate aim is to deactivate the majority of synths by unleashing power surges via their charging systems, followed by coordinated anti-synth groups attacks on the surviving synths. Thwarting their plans and with the televised sacrifice of one of the original conscious synths, humans and synths march on parliament for the sake of unity and peace.

The above is fiction, of course, but it has drawn on the culture and history of humanity to know how humans may react to robots. In this case, the majority of synths carry out domestic tasks, are almost too friendly, trusting, and docile. And yet, they are still resented and even more so for looking exactly like us except for bright green or blue eyes (and later orange). The show touches on our fears of being supplanted at work or even at home by living dolls. There are views on human rights for synths, punishments for crimes against synths, and people who want to be synths.

Among the drama of the show was the unstated parallels with our society. Without being explicit about it, the synths were treated like minorities are in today's society. It wasn't too long ago that ethnic minorities were looked upon as less than human. They were human robots toiling away for the colonial masters. Now the synths are the new domestic servants, the low-skilled workers, and the slaves. The synths may look like us, but the 'uncanny valley' factor creeped the humans out, and they couldn't see the synths for what they were: our mirror. And they didn't like or accept the reflection.

By its very name, it seems Musk wants Optimus to be our mirror, an optimistic reflection of humanity doing the work we do not want to do. It may not matter if Optimus is functionally successful or not; what may matter more is its looks. Roboticists around the world may be trying to help society by creating versions of humans to help us, but as per the discussions on form following function, we saw how it was more important to get the task detail right over the physical appearance of the robot.

However, if Humans are anything to go by, a humanoid-looking robot may not be so welcome. So, Optimus may have to be less than humanoid: Sub-Optimus, as Musk (perhaps presciently) has called it.

Computer-generated imagery (CGI) may be all the rage in movies, allowing for spectacular stunts, weird creatures, and recreating or de-aging actors. But what if the actors were autonomous humanoid robots wearing the copied faces of deceased actors, or were built to be out-sized creatures or performed stunts? What if they created their own original movies, music, and art? And as in Star Trek: Picard or Blade Runner, will there be future autonomous humanoid robots who would not even be aware they were artificial lifeforms? Will it matter? The idea of robots being the playthings of Man has existed for centuries going back to the simple automatons of ancient Greece.

The idea that robots could start to become self-aware is a well-explored theme. And the idea that humanity itself could become the pawn of the sentient machine is now almost a reality. Combine these concepts and you enter the dystopian show that is Westworld.

This may be where Optimus and other robots belong: out of sight and mind within their own world, which humans visit to interact and play. Of course, in the fictional world of the sci-fi, neo-Western drama Westworld, things go wrong. Westworld is a technologically advanced Wild-West-inspired theme park created by Delos Inc. in the 2050s, where the rich human guests get to live out their wildest dreams among the androids, known as hosts. In this often-violent world, the hosts cannot harm humans and are the victims of gruesome and/or sexual acts, all to serve the humans' fantasies. After each plotted storyline is complete, the hosts have their memories erased and their bodies repaired, ready for the next bunch of guests. An important point is made by Delos that the hosts cannot feel pain as they are just machines. However, after a program update is introduced, some of the hosts begin to experience consciousness and remember their previous "lives". They set out to protect themselves with the use of violence and find a way to survive.

Some of the hosts escape into the real world where they discover humanity is itself controlled by a global AI program. There's also the discovery that Delos has been using Westworld as a testbed to create a human immortality experiment—recreating and uploading saved human consciousness into an android. Seeing the dangers that the AI presents to both them and humanity, a host faction helps destroy the AI.

The series pushes the boundaries of what is the stuff of life. If an autonomous humanoid robot evolves beyond its programming and becomes conscious, is it alive?

Psychologist Julian Jaynes studies the mind and proposes that there are two separate minds—one that gives instructions, and another that performs them (the bicameral mind). Jaynes argues that consciousness results from the breaking down of the wall between the two sides, thereby subjecting the conscious being to new stimuli. If that can happen in humans and be naturally occurring in robots, then by definition they are conscious and alive. But will such an explanation be accepted by politicians, scientists, theologians, and the public? Does artificial creation really matter in the meaning of life?

First off, will Optimus and other humanoid robots find a

place in society, or will they find themselves confined to communities where humans have to visit them? If you looked out your window now to see autonomous robots going about their business, how would you feel and react?

It may seem that roboticists and their public relations and marketing teams will have a lot of work to do after presenting their robots to the world. Autonomous humanoid robots may be technology, but their impact will go a lot further than the latest Apple product, 3D printer, or autonomous drone. It may just be that humans are enticed to interact with robots within special parks or communities in order to trial the experience before ordering them for work or domestic settings.

Secondly, if consciousness were to occur, then as argued earlier in Chapter 4's Robot Laws and ethics, scientists and politicians must anticipate the future now and draw up provisional legislation for such a contingency. There could be nothing worse than being caught off-guard by a sentient robot awakening to a world that does not understand its new state of mind and place in the world. It should be welcomed and treated with the dignity it deserves. Science fact can be stranger than science fiction, but to some futurists, the convergence of humanity and robots is already underway.

"If you can't beat them, join them." With the alleged

robopocalypse on the way and with a chance to live forever, why not live out your days in the body of a humanoid robot? We've seen how Devanthro is fashioning Robodies so that one day, through teleoperation we can experience life through a robot's body. But what if we went a step further and your consciousness was uploaded to a humanoid robot body and you could live for as long as the uploads were good for new artificial bodies or other suitable digital systems?

That is the premise of the Singularity. In his first book The Age of Spiritual Machines (1999), futurist and inventor Ray Kurzweil discussed the potential for computers to surpass the extent of human intelligence. And with that ever-increasing technological development, his follow-up book The Singularity is Near (2005) investigated how humans could transcend their biology and merge with technology. The theory behind this is GRAIN (Genetics, Robotics, Artificial Intelligence, and Nanotechnology). In a simplified explanation, each of those elements would need to reach such a technological synergy that detailed scanning and transference of a human brain into a robot body or digital system like a computer would be possible. Of course, the Singularity movement is not without its critics, but we could extend life, encounter new environments, and be practically immortal.

Most people would have encountered the term or the idea of the Singularity in many sci-fi shows, and recently in two movies: Transcendence (2014) and Singularity (2017). Both movies have scientists who wind up uploaded into their sentient machines for various reasons. While one scientist's philanthropic goals are misunderstood and he is destroyed, the other becomes the enemy of humanity trying to destroy them with robots. Both movies had mixed to negative reviews, but for Transcendence, a critic from The (U.K.) Independent newspaper thought that the movie was a "fairly well-judged path between paranoia and technological utopianism," while AI researcher Stuart Russell had a serious message for his fellow researchers in that, "AI researchers must, like nuclear physicists and genetic engineers before them, take seriously the possibility that their research might actually succeed and do their utmost to ensure that their work benefits rather than endangers their own species." Both the movie review-criticisms are interesting takes from a fictional movie. There's the doom and gloom, but enough positiveness and hope for people to envision what the future could be like with the technology to transfer one's mind to another medium.

If this was possible in the future, and Optimus was found to be a viable receptacle for the human mind, how many would take the chance? Biologists are always hedging their bets on how long a human lifespan can last with the technologies at their disposal. Robots could put the biologists out of work with humans eschewing the natural aging process. And with autonomous humanoid robots, sentient or non-sentient, and with humans living within robot bodies living and working together, this would redefine what it means to be human. And where to next?

Space exploration

We all know Elon Musk is keen on Mars missions, with Starships lined up to start delivering a million people to Mars in the future. His first thoughts on a mission to Mars was with mice after he had attended a Mars Society conference in 2001. And after many years with SpaceX in full swing, will we see Optimus venture into space alongside NASA's Valkyrie or Pal Robotics' Talos? On the trip out, Optimus could be stowed away as many of the ship's systems would be automated with non-humanoid assistance required. Upon planetfall, Optimus may have the physical parameters necessary to carry out short planet-side exploratory missions into hazardous areas. But in the form following function mode, NASA and other space agencies will prefer tracked, wheeled, or quadruped robots over a humanoid robot for long-range exploring. Optimus would be ideal for the on-site building of the base, cleaning, maintenance, etc.

For the first missions to another solar system like Alpha Centauri around 4 light-years away, AHRs would probably be the best initial crew to send, depending on what percentage of the speed of light we could travel. If the spaceship could travel at 10 percent the speed of light then a one-way trip would take around 44 years. There would be no need to return the robots, who could set up communications and exploratory bases before humans arrived. With AHRs on another planet, they would be fitted with sensors to test the physical nature of the new world on a human body. The data would be invaluable once communicated back to Earth. And if on the remote chance there was intelligent life on the planet, the robots would also give a sense of a human's form to the aliens and be able to interact accordingly.

But first, we have to deal with life on Earth and humanity's interactions with Optimus. Musk has already contended that Optimus will be the future of Tesla. He is staking the future of his reputation and company on this. With robotics technology and human ingenuity accelerating, and with Musk fully determined to deliver, Optimus looks primed to be a success, whenever it arrives.

Chapter 8:
Reasons to be Optimistic for the future

"The robots of our youths and of our imaginations don't have to have millions of dollars of incredibly sophisticated hardware and software in them. They can be relatively lightweight dumb devices so long as they're connected to the power of the cloud."

-ALEC ROSS (American technology expert and Senior Advisor to the Secretary of State during the Obama administration)

Optimus is usering in a new generation of autonomous humanoid robots that will change the world forever, handling tasks humans don't want to do, and generally being a good safe friend to humans.

Is there an economic revolution to look forward to? With the anticipated role of Optimus and other humanoid robots to take over human jobs, there would certainly seem to be a downside to robots in the workplace. But we have seen that with careful planning, technological unemployment need not be the end of human work. Universal basic pay schemes and robot taxes can be established to alleviate the worst of the expected mass job losses. And when one type of industry changes another will take its place.

Humans may choose more creative and service work that robots would not be able to carry out. Our relationship with Optimus must be synergistic. If the humanoid robots are doing the work we do not want to do, most likely low-skilled monotonous work, then let them. People can re-skill or upskill, raising the standards of education, salaries, and lifestyles for many. What these jobs will be, we cannot be certain, as not all will be geared toward technology to keep the robots coming. But niche, freelance, temp work, creative work, etc, will be the dominant human industries to be reinvented by humans for humans.

We will need to be prepared for and address the challenges, risks, and dangers that will arise with a new technology that could one day outsmart and outpace us. We need backstops to prevent humanoid robots like Optimus from unplugging humans from the world. The AI brigade will have to safeguard us from sentient robots, realizing that it is of an order above us and that humanity is but a virus on its evolutionary path.

Musk wants a friendly Optimus, not a Terminator, but how will his ethical dilemma be resolved? In any case, we must accept it is not the robot's fault. We built Optimus in our image and it will have the same reasoning capacity as humans: whether to be merciful or vengeful. It is up to humans to show the way, not for the robots to rely on programming. Nurture the good, and good nature should follow.

The autonomous humanoid robot race is truly underway with Japan leading the global pack. It is a phenomenal industry and could be very financially rewarding, but roboticists and AI specialists are working toward a different kind of reward: that of creating a true sentient robot that can really help humanity, especially the elderly. It's a noble cause, but the selected dozen or so companies described in Chapter 5, working on their various humanoid robots, are plowing lonely lanes in a field crowded with non-humanoid robots which can do the same roles for less money and more efficiency. Companies will have to assess why they are building a humanoid robot that may look great and interact with humans, but also creep them out with their uncanny human looks. To avoid the 'yuck' factor, Optimus may have to be as mentioned above, Sub-Optimus, to remind us that robots are not us in disguise.

Even the most conservative criticism is harsh and the brotherhood of roboticists seems to be turning their backs to Tesla's efforts rather than banding together to offer encouragement. But Musk will persevere and Optimus could be the robot that does disrupt the industry, driving down production and sales costs. In the end, it is not Elon Musk's and Tesla's job to silence critics; it is to deliver on their promise to develop Optimus to its full potential. Critics be damned, Optimus is on the way.

The future for Optimus is far from laid out, but we can infer from other robotics programs and even fictional media, where Optimus may end up. No doubt, Optimus will have profound implications for our society and our economy, and force us to question our beliefs about who we are and how we live our lives.

The future with autonomous humanoid robots looks very bright and prosperous, but not without its risks for both robots and humans. Optimus will not be born into a vacuum. Its vacant slate will have to be filled with the best of humanity.

For even as Optimus helps humanity, it will also surely accompany us as we venture forward into the unknown.

References

Ackerman, E. (2012, March 27). Meka and UT Austin Developing "Hyper-Agile" Bipedal Robot. IEEE Spectrum.

https://spectrum.ieee.org/meka-and-ut-austin-developing-hyperagile-bipedal-robot

Ackerman, E., & Guizzo, E. (2013, December 4). Google Acquires Seven Robot Companies, Wants Big Role in Robotics. IEEE Spectrum.

https://spectrum.ieee.org/google-acquisition-seven-robotics-companies

Agility Robotics. (n.d.). Agility Robotics. https://agilityrobotics.com/

Ameca. (n.d.). Engineered Arts.https://www.engineeredarts.co.uk/robot/ameca/

ASIMO by Honda | The World's Most Advanced Humanoid Robot. (2019). Honda.com. https://asimo.honda.com/

Bellan, R. (2022, April 21). Musk says Optimus robot will be worth more than Tesla's car business. TechCrunch.

https://techcrunch.com/2022/04/20/musk-says-robot-aimed-for-2023-will-be-worth-more-than-teslas-car-business/

Boston Dynamics. (2019). Boston Dynamics is changing your idea of what robots can do. | Boston Dynamics. Bostondynamics.com.

https://www.bostondynamics.com/

Countries That Have Tried Universal Basic Income 2020. (n.d.).

Worldpopulationreview.com. https://worldpopulationreview.com/country-rankings/countries-with-universal-basic-income

Delaney, K. J. (2018, September 23). Bill Gates: the robot that takes your job should pay taxes. Quartz; Quartz.

https://qz.com/911968/bill-gates-the-robot-that-takes-your-job-should-pay-taxes/

Devanthro – the Robody Company. (n.d.). Retrieved May 8, 2022, from

https://devanthro.com/

Diamandis, P. H., & Kotler, S. (2017). Bold: how to go big, create wealth and impact the world. Simon & Schuster.

Elon Musk REVEALS Tesla Bot (full presentation). (2021, August 19).

www.youtube.com
https://www.youtube.com/watch?v=HUP6Z5voiS8&ab_

channel=CNETHighlights

Elon Musk Unveils Humanoid Robot to Take Over "Boring" Work. (2021, August 20). Bloomberg.com. https://www.bloomberg.com/news/articles/2021-08-20/tesla-s-musk-unveils-humanoid-robot-to-take-over-boring-work

Graham, J. (n.d.). Erica, the humanoid robot, is chatty but still has a lot to learn. USA TODAY. https://eu.usatoday.com/story/tech/talkingtech/2018/02/19/erica-humanoid-robot-chatty-but-still-has-lot-learn/352281002/

Hanson Robotics Limited. (2016). Hanson Robotics. https://www.hansonrobotics.com/

Heater, B. (2021, September 1). Agility Robotics' Digit gets a warehouse gig.

TechCrunch. https://techcrunch.com/2021/09/01/agility-robotics-digit-gets-a-warehouse-gig/?guccounter=1

Heater, B. (2022, March 7). Robotics founders: Build your pitch deck around problem-solving, not technology. TechCrunch.

https://techcrunch.com/2022/03/07/robotics-founders-build-your-pitch-deck-around-problem-solving-not-technology/?cx_testId=6&cx_testVariant=cx_1&cx_artPos=3#cxrecs_s

Home. (n.d.). PROMOBOT. Retrieved May 8, 2022,

from https://promo-bot.ai/

Home - iCub - IIT. (n.d.). Icub.iit.it. https://icub.iit.it/

Hornyak, T. (2019, October 31). Insanely humanlike androids have entered the workplace and soon may take your job. CNBC; CNBC. https://www.cnbc.com/2019/10/31/human-like-androids-have-entered-the-workplace-and-may-take-your-job.html

Hsu, J. (2012, April 30). Control AI before it controls us, expert says - Technology & science - Innovation - msnbc.com. Web.archive.org. https://web.archive.org/web/20120430093909/http://www.msnbc.msn.com/id/46590591/ns/technology_and_science-innovation

iCub. (2022, February 18). Wikipedia. https://en.m.wikipedia.org/wiki/ICub

iCub - ROBOTS: Your Guide to the World of Robotics. (n.d.). Robots.ieee.org. Retrieved May 8, 2022, from https://robots.ieee.org/robots/icub/

Inside the Life of People Married to Robots. (2020, February 18). Buzzworthy. https://www.buzzworthy.com/meet-men-married-robots/

jinn-Bot - independent Swiss research and development firm, creating humanoid robots. (n.d.). Jinn-Bot.com. Retrieved May 8, 2022, from https://jinn-bot.com/

Kurzweil, R. (2005). The singularity is near: when humans transcend biology.

Duckworth.

Leju – The Leading Company of Humanoid Robots. (n.d.). LejuRobot. Retrieved May 8, 2022, from http://www.lejurobot.com/

Leswing, K. (2021, August 20). Elon Musk says Tesla will build a humanoid robot prototype by next year. CNBC.

https://www.cnbc.com/2021/08/19/elon-musk-teases-tesla-bot-humanoid-robot-for-repetitive-tasks.html

List of Humans episodes. (2022, April 25). Wikipedia.

https://en.wikipedia.org/wiki/List_of_Humans_episodes

MacIntyre, D. (2021, November 24). Panorama - The Electric Car Revolution: Winners and Losers. BBC One.

Marshall, A. (2019, November 23). Why the Tesla Cybertruck Looks So Weird. WIRED.https://www.wired.com/story/why-tesla-cybertruck-looks-weird/

Nations, U. (n.d.). At UN, robot Sophia joins meeting on artificial intelligence and sustainable development. United Nations.

https://www.un.org/en/desa/un-robot-sophia-joins-meeting-artificial-intelligence-and-sustainable-development

Nwodo, B., Jhaveri, N., Gomila, J., Goyal, R., Shagivildanovna, L., Sharma, D., & Singh, A. (n.d.). Agility Robotics - Wiki. Golden. Retrieved May 8, 2022, from https://golden.com/wiki/Agility_Robotics-JNZJAX3

O'Kane, S. (2021, August 19). Elon Musk says Tesla is working on humanoid robots. The Verge.

https://www.theverge.com/2021/8/19/22633514/tesla-robot-prototype-elon-musk-humanoid-ai-day

palrobot. (n.d.). PAL Robotics: Leading company in service robotics. PAL Robotics. https://pal-robotics.com/

Price, A. (2010, October 13). Automation Insurance: Robots Are Replacing Middle Class Jobs - Business - GOOD. Web.archive.org.

https://web.archive.org/web/20120419153141/http://www.good.is/post/automation-insurance-robots-are-replacing-middle-class-jobs/

Robots Today and Tomorrow: IFR Presents the 2007 World Robotics Statistics Survey. (2007, October 27). Web.archive.org.

https://web.archive.org/web/20080205041924/http://www.robots.com/blog.php?tag=48

Roboy. (2021, July 7). Wikipedia. https://en.wikipedia.org/wiki/Roboy

Russell, J. (2018, November 15). Google is closing its Schaft robotics unit after failing to find a buyer. TechCrunch.

https://techcrunch.com/2018/11/14/google-is-closing-its-schaft-robotics-unit/

Shead, S. (2022, April 8). Elon Musk says production of Tesla's Optimus humanoid robot could start next year. CNBC.

https://www.cnbc.com/2022/04/08/elon-musk-says-tesla-is-aiming-to-start-production-on-optimus-next-year.html

Siddiqui, F. (2021, August 20). Tesla says it is building a "friendly" robot that will perform menial tasks, won't fight back. Washington Post.

https://www.washingtonpost.com/technology/2021/08/19/tesla-ai-day-robot/

Simonite, T. (2008 11). Six ways to build robots that do humans no harm - tech - 18 November 2008 - New Scientist. Web.archive.org.

https://web.archive.org/web/20120707060741/http://www.newscientist.com/article/dn16068-six-ways-to-build-robots-that-do-humans-no-harm.html

Singularity (2017 film). (2022, April 25). Wikipedia.

https://en.wikipedia.org/wiki/Singularity_(2017_film)

SoftBank Robotics | Humanoid and programmable robots. (n.d.).

www.softbankrobotics.com. https://www.softbankrobotics.com/emea/en/index

Taylor, C. (2022, April 11). Three reasons universal basic income pilots haven't led to policy change – despite their success. The Conversation.

https://theconversation.com/three-reasons-universal-basic-income-pilots-havent-led-to-policy-change-despite-their-success-180062

TechCrunch's robotics newsletter, Actuator. (n.d.). TechCrunch. Retrieved May 8, 2022, from

https://techcrunch.com/2022/04/28/optimus-not-ready-for-prime-time/

Tesla Bot. (2022, April 9). Wikipedia. https://en.wikipedia.org/wiki/Tesla_Bot

Tesla Bot: AI-controlled humanoid robot revealed. (n.d.). Motor Authority.

https://www.motorauthority.com/news/1133301_tesla-bot-ai-controlled-humanoid-robot-revealed

Tesla Promised a Robot. Was It Just a Recruiting Pitch? (n.d.). Wired.

https://www.wired.com/story/tesla-promised-robot-recruiting-pitch/

Tonkin, S. (2021, December 3). "World's most advanced" humanoid robot is unveiled in a UK lab. Mail Online.

https://www.dailymail.co.uk/sciencetech/article-10270925/Worlds-advanced-humanoid-robot-unveiled-UK-lab.html?ito=email_share_article-masthead

UBTECH Robotics. (n.d.). UBTECH Robotics. Retrieved May 8, 2022, from https://starwars.ubtrobot.com/?ls=en

Vance, A. (2016). Elon Musk: how the billionaire CEO of SpaceX and Tesla is shaping our future. Virgin Books.

Vega, N. (2022, April 21). Elon Musk says Tesla's humanoid Optimus robot "will be worth more than the car business." CNBC.

https://www.cnbc.com/2022/04/21/elon-musk-says-optimus-robot-will-be-worth-more-than-tesla.html

Vincent, J. (2021, August 20). Don't overthink it: Elon Musk's Tesla Bot is a joke. The Verge. https://www.theverge.com/2021/8/20/22633958/tesla-bot-elon-musk-ai-day

Watanabe, H., & Negishi, M. (2007, December 2). Japan's robots slug it out to be world champ. Reuters.

https://www.reuters.com/article/technologyNews/idUST32811820071202

Wikipedia Contributors. (2019a, April 7). SolarCity. Wikipedia; Wikimedia Foundation.

https://en.wikipedia.org/wiki/SolarCity

Wikipedia Contributors. (2019b, April 7). Technological unemployment. Wikipedia; Wikimedia Foundation.

https://en.wikipedia.org/wiki/Technological_unemployment

Wikipedia Contributors. (2019c, April 30). Transcendence (2014 film). Wikipedia; Wikimedia Foundation.

https://en.wikipedia.org/wiki/Transcendence_(2014_film)

Wikipedia Contributors. (2019c, May 8). Tesla. Wikipedia; Wikimedia Foundation.
https://en.wikipedia.org/wiki/Tesla

Wikipedia Contributors. (2019d, October 26). Humans (TV series). Wikipedia; Wikimedia Foundation.
https://en.wikipedia.org/wiki/Humans_(TV_series)

Wikipedia Contributors. (2019e, November 7). Electromagnetic pulse. Wikipedia; Wikimedia Foundation.
https://en.wikipedia.org/wiki/Electromagnetic_pulse

Wikipedia Contributors. (2019f, December 2). Westworld (TV series). Wikipedia; Wikimedia Foundation.
https://en.wikipedia.org/wiki/Westworld_(TV_series)

www.ingramcontent.com/pod-product-compliance
Lightning Source LLC
Chambersburg PA
CBHW052355220526
45465CB00003BA/1111